ACS SYMPOSIUM SERIES 249

The Chemistry of Combustion Processes

Thompson M. Sloane, EDITOR

General Motors Research Laboratories

Based on a symposium sponsored by
the Division of Industrial
and Engineering Chemistry
at the 185th Meeting
of the American Chemical Society,
Seattle, Washington,
March 20–25, 1983

American Chemical Society, Washington, D.C. 1984

69866673

SEP/AE
CHEM

Library of Congress Cataloging in Publication Data

The chemistry of combustion processes.
  (ACS symposium series, ISSN 0097-6156; 249)

  Includes bibliographies and indexes.

  1. Combustion—Congresses.

  I. Sloane, Thompson M., 1945-   . II. American
Chemical Society. Division of Industrial and
Engineering Chemistry. III. American Chemical
Society. Meeting (185th: 1983: Seattle, Wash.)
IV. Series.

QD516.C537   1984   541.3'61   84-2816
ISBN 0-8412-0834-4

FOREWORD

The ACS SYMPOSIUM SERIES was founded in 1974 to provide a medium for publishing symposia quickly in book form. The format of the Series parallels that of the continuing ADVANCES IN CHEMISTRY SERIES except that in order to save time the papers are not typeset but are reproduced as they are submitted by the authors in camera-ready form. Papers are reviewed under the supervision of the Editors with the assistance of the Series Advisory Board and are selected to maintain the integrity of the symposia; however, verbatim reproductions of previously published papers are not accepted. Both reviews and reports of research are acceptable since symposia may embrace both types of presentation.

CONTENTS

PREFACE

IN RECENT YEARS, combustion science has undergone a dramatic transformation as modern analytical tools have been applied to the study of combustion systems. Techniques involving molecular beams and lasers, which have been so successful at breaking new ground in the fields of spectroscopy and chemical reaction kinetics and dynamics, are now yielding information in increasingly microscopic detail about the physics and chemistry of combustion processes. These new techniques, particularly laser techniques, have gone beyond the demonstration stage and have made significant contributions to our knowledge.

This volume focuses on the understanding of chemical aspects of combustion processes that has been achieved with these new experimental methods and with the concurrent increase in theoretical work. The significance of these topics in current combustion science is due principally to their role in the strongly linked areas of energy conservation, pollutant emissions, and safety. For example, the subject of detonations is of primary importance in the transportation safety of natural gas in both the liquid and gaseous state. New methods of ignition may yield more efficient combustion processes in automobile engines. A better understanding of the chemistry of fuel nitrogen conversion to nitric oxide will aid in reducing nitric oxide emissions from coal-fired combustors and from combustors utilizing heavy fuels made from coal and shale. A determination of the mechanism of soot formation and destruction will help to make the highly efficient diesel engine more environmentally acceptable in transportation vehicles and should also aid in reducing soot emissions from combustors burning coal and heavy liquid fuels.

I would like to thank the authors and speakers for their contributions that made this symposium such a worthwhile endeavor. Thanks are also due to the publishers for providing the means to record the proceedings of this symposium.

THOMPSON M. SLOANE
General Motors Research Laboratories
Warren, Michigan

December 1983

SOOT

Role of C$_4$ Hydrocarbons in Aromatic Species Formation in Aliphatic Flames

J. A. COLE[1], J. D. BITTNER[2], J. P. LONGWELL, and J. B. HOWARD

Department of Chemical Engineering, Massachusetts Institute of Technology, Cambridge, MA 02139

The free-radical addition reaction expressed as
1,3-butadienyl + acetylene → benzene + H-atom
is shown to account for benzene formation in the
preheat region of a low-pressure, near-sooting,
premixed 1,3-butadiene flame using an estimated
rate constant, Log k = 8.5 - 3.7/2.3 RT ℓ/(mol·s),
and measured species concentration profiles. Using
similarly estimated rate constants with activation
energies of 1.8, 0.6, and 3.7 kcal/mol it is shown
that 1,3-butadienyl addition to C_4H_2, C_4H_4, and
C_3H_4 also accounts for the formation of phenylacetylene and styrene but not toluene. Other reaction
mechanisms involving C-4 hydrocarbons also are considered but are too slow in this flame. Concentration and molar flux profiles obtained by molecular-
beam sampling with on-line mass spectrometry are
presented for 31 species.

Aromatic hydrocarbons are known to be important in soot formation
in flames. The aromatic structure may abet molecular growth leading to PAH and soot formation through its ability to stabilize
radicals formed from addition of aromatic radicals to unsaturated
aliphatics such as acetylenic species ([1],[2]). Accordingly, both
aromatics and unsaturated aliphatics would be important for growth
processes. Both types of species are prevalent in the flame zone
where growth occurs. Aromatic structures with unsaturated side
chains also are observed there ([1],[3]).

[1]Current address: Energy and Environmental Research Corporation, 18 Mason, Irvine, CA 92714
[2]Current address: Cabot Corporation, Concord Road, Billerica, MA 01821

0097-6156/84/0249-0003$06.00/0
© 1984 American Chemical Society

In aliphatic flames aromatic rings must be formed from non-aromatic precursors, but an accepted mechanism for this critical step has not appeared in the literature. Attempts to derive mechanisms for benzene formation in flames have suffered primarily from the lack of pertinent kinetic or thermodynamic data. Furthermore, we have found no literature wherein formation rates of benzene, or other aromatics, predicted from a mechanism are compared with rates measured in a flame.

We will not attempt to review here the many mechanisms which have been proposed to account for aromatic formation in aliphatic flames. Suffice it to say that these fall basically into three categories: ionic mechanisms; concerted, pericyclic mechanisms; and free radical mechanisms.

The reaction kinetics of different hydrocarbon ions in flames are not sufficiently understood to permit testing of proposed ion-molecule reactions leading to aromatic formation. This area is of interest for further study, especially of growth mechanisms for larger aromatic species and soot. With regard to pericyclic mechanisms such as Diels-Alder reactions, observations made in butadiene flames have led to proposals that butadiene reacting with other olefins might be responsible for aromatic formation (4-6). Flame and pyrolysis studies, however, have shown no evidence of Diels-Alder reactions. As described below, we have compared known reaction rates with measured formation rates in butadiene flames and found Diels-Alder reactions to be too slow to be significant (7,8).

Nevertheless, a link between four-carbon species and PAH formation would be consistent with the prominence of C-4 hydrocarbons in flames where PAH or soot are being formed. In particular, 1-buten-3-yne (C_4H_4; vinylacetylene) concentration profiles mimic those of benzene and PAH in many fuel-rich flames (9).

In this work the net formation rates of benzene, toluene, phenylacetylene, and styrene, and concentrations of possible precursors, were determined as a function of distance from the burner in the primary reaction zone of a low-pressure, near-sooting 1,3-butadiene-oxygen-4% argon flame. These rates are compared with estimated rates for several reaction mechanisms.

Experimental

The apparatus is a low-pressure flat-flame burner with a molecular-beam sampling instrument having a quartz probe and an on-line quadrupole mass-spectrometer (Figure 1). The design features and measurement techniques are the same as those described elsewhere (1,10).

A near-sooting laminar premixed flat flame was produced at a burner chamber pressure of 2.67 kPa (20 torr) with 52.1 normal $cm^3 \cdot s^{-1}$ (0.1 MPa, 298 K) of feed gas consisting of 29.5 mol% 1,3-butadiene, 67.5 mol% oxygen, and 3.0 mol% argon, corresponding to a fuel-equivalence ratio, ϕ, of 2.4 and a cold-gas velocity of

$0.5 \text{ m} \cdot \text{s}^{-1}$ at 298 K. At this pressure and cold-gas velocity the sooting limit was found at $\phi = 2.46$.

All of the gases used were more than 99% pure and were used as supplied by the manufacturers.

Individual species were identified by mass and ionization potential. Their mole fractions were measured at the centerline of the flame as a function of distance from the burner as described elsewhere (1,10).

Estimated probable errors for species mole fraction are shown in Table I below. The mole fractions of the largest hydrocarbon

Table 1.   Estimated Probable Errors for Species Mole-Fraction Measurements

Species	Estimated Probable Error
Major Species	± 3%
H_2O	± 8%
C_2H_2	+12%
Hydroxyl	+17%
H-atom	+19%
Other radicals	+50%
Species larger than Benzene	factor of two

species were determined relative to each other with greater accuracy than the errors associated with their absolute mole fractions would suggest.

The mole fraction profiles were numerically smoothed and differentiated in order to determine the species' molar fluxes. This process was repeated on the flux profiles in order to provide net molar rates of formation or destruction for each species. The numerical techniques have been described previously (1,2,7,10). Probable errors associated with fluxes and reaction rates so determined cannot be stated explicitly because the perturbation of the sampling probe on the one-dimensional flame assumption has not been assessed quantitatively.

Nevertheless, the overall accuracy suggested by element balances is encouraging. The calculated total-carbon mass flux agrees to within 30% throughout the flame with the value determined from the measured fuel feedrate. The agreement is to within 5% in the region of the flame considered in this work.

The gas temperature profiles used for this analysis were calculated from the assumption of equilibrium in the reaction

$$H \cdot + H_2O \rightleftarrows H_2 + OH \tag{1}$$

This reaction appears to be equilibrated beyond the primary reaction zone (region of total fuel and/or oxygen consumption) in this

flame and elsewhere (11). The temperature was calculated by comparing the equilibrium equation

$$K = \frac{X_H \ X_{H_2O}}{X_{H_2} \ X_{OH}}$$

with the known standard Gibbs energy charge, $\Delta G°$, for Reaction (1)
as a function of temperature (12). In this case the equilibrium
constant, K, is related to $\Delta G°$ by

$$\Delta G° = -RT \ln K$$

The resulting temperature profile is shown in Figure 2. The temperature in the primary reaction zone was estimated by linear extrapolation to 300 K at the burner. Propagation of error calculations suggests a maximum probable error of ±150 K.

Experimental Results and Discussion

The mole-fraction and molar-flux profiles obtained for 31 species
are shown in Figures 3-8. Inspection of the flux profiles in
Figure 3 reveals little evidence of reaction up to about 5 mm from
the burner. Butadiene is consumed in the region 5-10 mm, although
primarily beyond 7 mm (Figure 3). Oxygen consumption, however,
does not become noticable until about 8.5 mm (Figure 4) and production of CO, CO_2, and H_2O occurs later still (Figures 3 and 5).
Nevertheless some oxidation occurs prior to 8.5 mm as is evidenced
by the fluxes of the oxygenated species in Figure 4. The species
C_3H_4O and C_4H_6O probably result from triplet (ground state) O-atom
attack on butadiene and allene (13,14). Therefore O-atom may play
a role in butadiene destruction. Such behavior would be consistent
with the observation that the fractional decomposition rate (s^{-1}) of
butadiene at the 1120 K position in this flame, where O atoms are
available by diffusion from the oxidation zone, is significantly
larger than that measured in the presence of O_2 at 1120 K in an
approximately isothermal flow reactor (15).
      Hydrocarbon species with one, two, or three carbon atoms are
represented in Figure 6; those with four carbon atoms in Figure 7.
The behavior and position of the C_4H_4 mole fraction profile (Figure 7) are strikingly similar to those of all the aromatic species
shown in Figure 8. In contrast, the profiles of diacetylene
(C_4H_2, Figure 7) and the polyacetylenes (C_6H_2 and C_8H_2, Figure 8)
are similar to those of acetylene (Figure 6).
      Figure 8 shows that benzene and the other single-ring aromat-
ics whose formation rates are studied here are formed mainly in the
region of butadiene consumption. Their maximum net rates of formation occur prior to 9.5 mm. Since significant O_2 consumption
begins at 8.5 mm and oxidation is evident prior to that, the question arises as to whether oxidative consumption of aromatics contributes significantly to their net reaction rates in the region
of interest here

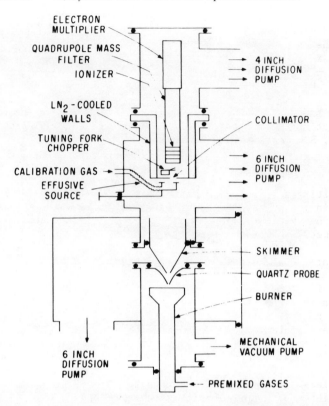

Figure 1.   Molecular beam mass spectrometer flame sampling apparatus.

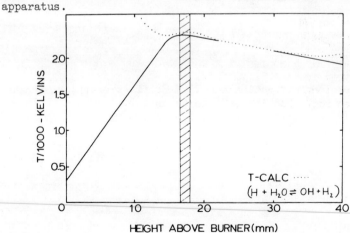

HEIGHT ABOVE BURNER(mm)

Figure 2.   Temperature profile in near-sooting. Points, calculated assuming equilibration of $H + H_2O \rightleftharpoons OH + H_2$; and curve, estimated and used in data analysis and mechanism predictions. Conditions: $\phi = 2.4$; $C_4H_6/O_2/3.0\%Ar$ flame; cold gas velocity = 0.5 m/s; and pressure = 2.67 kPa.

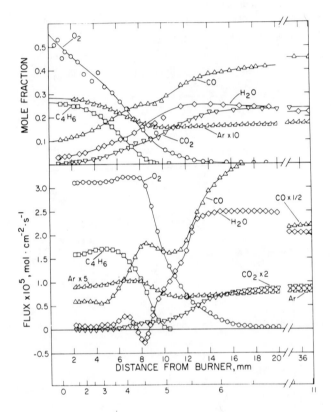

Figure 3.   Mole fraction and flux profiles of major species in near-sooting.   Conditions same as in Figure 2.

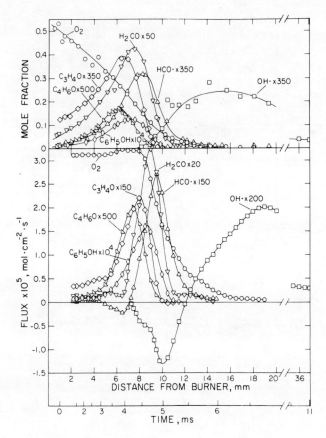

Figure 4.   Mole fraction and flux profiles of OH, O₂, and oxygenated hydrocarbons in near-sooting.   Conditions same as in Figure 2.

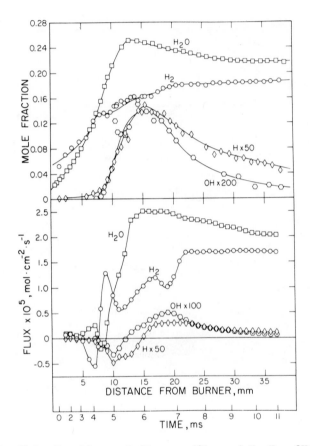

Figure 5. Mole fraction and flux profiles of H, H_2, OH, and H_2O in near-sooting. Conditions same as in Figure 2.

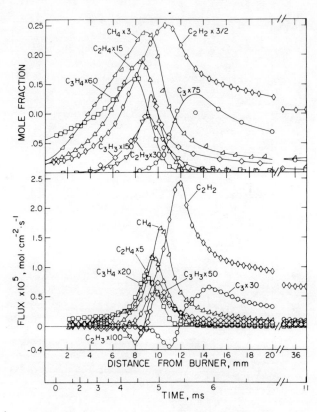

Figure 6. Mole fraction and flux profiles of one-, two-, and three-carbon species in near-sooting. Conditions same as in Figure 2.

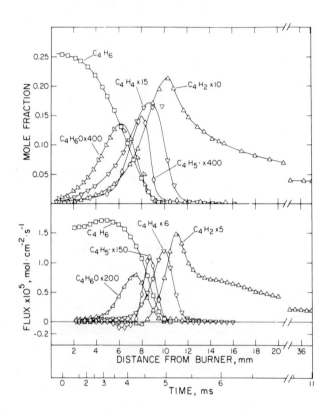

Figure 7. Mole fraction and flux profiles of four-carbon species in near-sooting. Conditions same as in Figure 2.

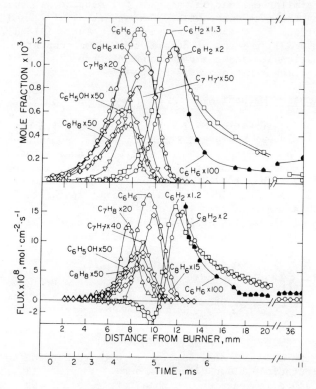

Figure 8.   Mole fraction and flux profiles of species with masses
between 74 and 104 amu in near-sooting.   Conditions same as in
Figure 2.

Using the rate constant of Tully et al. ([16]), the temperature
profile of Figure 2, and the measured benzene and hydroxyl mole
tions, the predicted rate of benzene destruction by OH is insignif-
icant compared to the measured net benzene formation rate until
after the latter passes through its maximum value at 8.5 mm. At
8.5 mm the predicted destruction of benzene by hydroxyl is one percent of the measured formation rate, increasing only to ten percent
at 9.3. A similar estimation was made for benzene destruction by
O-atom using the rate constant of Nicovich et al. ([17]) and reasonable estimations for O-atom mole fractions. This yielded values of
two percent and 17 percent at 8.5 mm and 9.3 mm respectively. The
lack of additional oxidation rate constants precluded similar calculations for the other aromatic species.

Benzene Formation Mechanisms

The formation of aromatics by reactions involving 1,3-butadiene was
considered. Several Diels-Alder reactions were considered, the
fastest of which is $C_4H_6 + C_2H_2$ owing to its large Arrhenius factor
and the high concentration of acetylene. However, the rate of this
reaction is less than 10^{-3} of the measured rate of benzene formation. Therefore Diels-Alder reactions do not appear to be significant for aromatics formation in this system.

Free radical mechanisms involving 1,3-butadiene also have been
considered. The reaction sequence

$$C_2H_3 \cdot + C_4H_6 \rightleftarrows C_6H_9 \cdot \tag{2}$$

$$C_6H_9 \cdot \rightleftarrows c\text{-}C_6H_9 \cdot \tag{3}$$

$$c\text{-}C_6H_9 \cdot \rightleftarrows c\text{-}C_6H_8 + H \cdot \tag{4}$$

$$c\text{-}C_6H_8 \rightarrow C_6H_6 + H_2 \tag{5}$$

provides a thermodynamically favorable route to benzene formation
in the 1,3-butadiene flame discussed here. The rate limiting step
in this sequence is Reaction 5 with a rate constant Log k_5 =
12.4-44/θ (s^{-1}), where θ = 2.3 RT kcal/mol ([18,19]). However, the
concentration of C_6H_8 in this flame is only a few percent of that
of benzene ([20]). This leaves Reaction 5 too slow by a factor of
10^5 to 10^6.

An alternative to Reaction 5 is the reaction sequence

$$c\text{-}C_6H_8 + R \cdot \rightleftarrows c\text{-}C_6H_7 \cdot + RH \tag{6}$$

$$c\text{-}C_6H_7 \cdot \rightleftarrows C_6H_6 + H \cdot \tag{7}$$

Reaction 6 is rate limiting here with a typical rate constant of
Log k_6 = 8.5-5.5/θ ($\ell \cdot mol^{-1} s^{-1}$) ([21]) for metathesis by hydrocarbon
radicals. H-abstraction by H-atom might be faster, with an Arrhen-

ius factor conceivably as large as 10^{11}. However, the predicted rate of benzene formation would still be less than a few percent of that which is observed.

In addition reaction sequences involving C_4H_4 as a benzene precursor were considered. Because of the required protonation of the number 3 carbon atom, however, all of these mechanisms proved to be too slow (7).

Radical addition of 1,3-butadienyl to acetylene provides a route to benzene formation through the sequence

$$1,3\text{-}C_4H_5\cdot + C_2H_2 \rightleftharpoons C_6H_7\cdot \qquad (8)$$

$$C_6H_7\cdot \rightleftharpoons c\text{-}C_6H_7\cdot \qquad (9)$$

$$c\text{-}C_6H_7\cdot \rightleftharpoons C_6H_6 + H\cdot \qquad (7)$$

The 1,3-butadienyl radical is primarily a by-product of butadiene pyrolysis in this system but results from vinyl addition to acetylene in flames of other aliphatic fuels. In aromatic flames 1,3-butadienyl may be produced by oxidative and pyrolytic decomposition of aromatic species, as suggested in a study of benzene flames (10).

Although Reaction 8 has not been studied explicitly, a rate constant of Log k_{10} = 8.5-3.7/θ ($\ell\cdot mol^{-1}s^{-1}$) was inferred from measured data for similar reactions (7). Comparison of this value with rate estimates for similar reactions (22) suggests that this actually may be a rather conservative figure. Reaction 8 is rate limiting in this sequence.

Figure 9 shows the agreement between the predicted benzene formation rate using this mechanism and the measured rate throughout the region of net benzene production. The measured rate is 5 to 10 times higher than the predicted rate. In view of the uncertainties encountered at each step in the analysis we feel that this level of agreement, coupled with the congruence of the formation rate profiles, provides evidence of the viability of this mechanism in the 1,3-butadiene flame.

Formation of Other Aromatic Species

Because other acetylenic species were also present in the butadiene flame, rate constants were estimated for the addition of three of them, C_3H_4 (propyne), C_4H_2 (butadiyne), and C_4H_4 (buten-3-yne) to 1,3-butadienyl (7). These addition reactions with subsequent cyclization of the adduct, followed by H-elimination (analogous to Reactions 8, 9, and 7) lead to formation of toluene, phenylacetylene, and styrene, respectively.

The rate constants shown below (Table II) were extrapolated from rate data using an Evans-Polanyi approach similar to that discussed recently by McMillen and Golden (23). The assumption of equal Arrhenius parameters is inherent in the estimation technique. Subsequent cyclization and H-elimination steps are affected negligibly

Table II.   Rate Constants for Addition of Acetylenic Species to
1,3-butadienyl at 800 K

Equation		Rate Constant, k, $\ell \cdot mol^{-1} \cdot s^{-1}$
$1,3\text{-}C_4H_5 \cdot + C_3H_4 \rightleftarrows C_7H_9 \cdot$	(10)	$10^{8.5-3.7/\theta}$
$1,3\text{-}C_4H_5 \cdot + C_4H_2 \rightleftarrows C_8H_7 \cdot$	(11)	$10^{8.5-1.8/\theta}$
$1,3\text{-}C_4H_5 \cdot + C_4H_4 \rightleftarrows C_8H_9 \cdot$	(12)	$10^{8.5-0.6/\theta}$

by the presence of side chains on the aromatic ring (24). Consequently, Reactions 10-12 are rate limiting for their respective
reaction sequences.

Figures 10-12 show the measured rates of formation of toluene,
phenylacetylene, and styrene compared with the rates predicted by
Reactions 10-12. Of these three, only toluene (Figure 10) shows a
serious disagreement between prediction and observation. The early
decline in the toluene formation rate (~ 7.5 mm) may be due to rapid
benzyl radical formation. A cursory examination of the C_7H_7 and
C_7H_8 flux profiles in Figure 8 supports this possibility. Nevertheless, the magnitude of discrepancy between measured and predicted rates suggests that some other mechanism is probably responsible
for toluene formation in this system.

Role of Vinylacetylene

The behavior of C_4H_4 relative to benzene and PAH has been observed
in other aliphatic flames, including those of methane (25,26),
acetylene (7,27), and ethylene (27), as well as benzene flames (1,
10). As an example, Figure 13 shows data for ethylene and acetylene flames extracted from the works of Crittenden (28) and Crittenden and Long (27). This correlation may be explained if 1,3-
butadienyl can be shown to be the primary precursor for formation
of C_4H_4, as well as PAH.

That C_4H_4 is indeed related to C_4H_5 is indicated by the present data as well as data from two other flames studied in the same
experimental system at the same burner velocity and pressure, name-
ly, a stoichiometric 1,3-butadiene flame (7) and a near-sooting
benzene flame (10). In all three cases, the ratio of the peak concentration of C_4H_5 to that of C_4H_4 is about 1:30 (e.g., Figure 7),
and the C_4H_5 peak occurs about 1 mm (or about 0.3 ms) before that
of C_4H_4. This behavior is indicative of an intermediate/product
relationship between C_4H_5 and C_4H_4. The kinetics of the relationship seems to vary little among these different flames.

In the present data, the net formation rate of C_4H_4 peaks exactly where the net rate of C_4H_5 consumption is maximum (9.3 mm),
and the former is about 10-times the latter. It is likely that
C_4H_4 is formed by Reaction (13). Equilibrium for this reaction

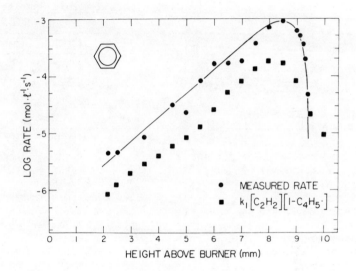

Figure 9. Comparison of measured net rates of benzene formation with predicted formation rates for near-sooting. Conditions same as in Figure 2.

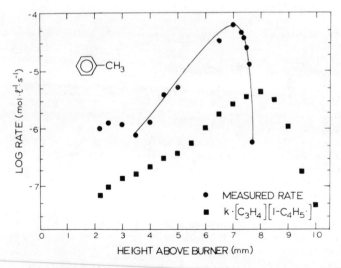

Figure 10. Comparison of measured net rates of toluene formation with predicted formation rates for near-sooting. Conditions same as in Figure 2.

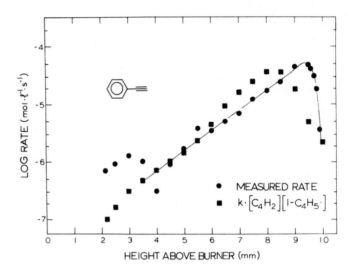

Figure 11. Comparison of measured net rates of phenylacetylene formation with predicted formation rates for near-sooting. Conditions same as in Figure 2.

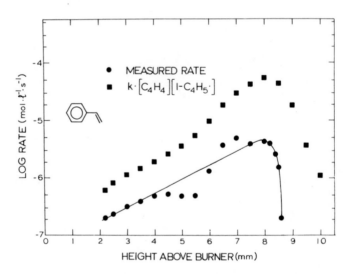

Figure 12. Comparison of measured net rates of styrene formation with predicted formation rates for near-sooting. Conditions same as in Figure 2.

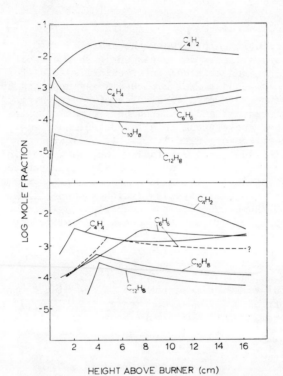

Figure 13. C_4H_2, C_4H_4, and PAH mole fraction profiles in (top) $C_2H_2/O_2/Ar$ flame and (bottom) $C_2H_4/O_2/Ar$ flame.   Data from Crittenden and Long (27) and Crittenden (28).   The dashed line for C_6H_6 in the lower figure suggests a discrepancy between the data presented in the two references.

$$1,3\text{-}C_4H_5 \overset{\rightarrow}{\leftarrow} C_4H_4 + H\cdot \tag{13}$$

favors C_4H_4 production throughout the region of PAH production (<9 mm). Furthermore, using an estimated rate constant (7) of Log $k_{13} = 13.8 - 41.7/\theta$ (s^{-1}) the predicted rate of C_4H_4 production from 1,3-butadienyl exceeds the measured rate by a factor of nearly 200 at 9.3 mm.

Conclusions

The observed relationship between the profiles of C_4H_4 and aromatics in hydrocarbon flames is consistent with the hypothesis that 1,3-butadienyl is the critical intermediate species for the formation of these compounds. The hypothesis is supported by considerable agreement between predicted rates of addition reactions of 1,3-butadienyl radical and measured net formation rates of benzene, phenylacetylene, and styrene in a 1,3-butadiene flame. However, the need for better rate data is clearly indicated.

The hypothesis that 1,3-butadiene may be responsible for benzene production is inconsistent with the data presented here. However, this conclusion is based on the low measured mole fractions of the intermediate species cyclohexadiene (C_6H_8) which may be peculiar to these experimental conditions. Therefore, this conclusion does not rule out the possible significance of 1,3-butadiene in other combustion systems.

Acknowledgments

We are grateful to Exxon Research and Engineering Company for financial support of this research through the EXXON/MIT Combustion Research Program, and to Phillip R. Westmoreland for many valuable discussions.

Literature Cited

1. Bittner, J. D.; Howard, J. B., 18th Symp. (Intl.) on Combustion, The Combustion Institute, 1981, p. 1105.
2. Bittner, J. D.; Howard, J. B., in "Particulate Carbon-Formation During Combustion"; Siegla, D. C.; Smith, G. W., Eds.; Plenum: New York, 1981; pp. 109-137.
3. Homann, K. H.; Wagner, H. Gg., 11th Symp. (Intl.) on Combustion, The Combustion Institute, 1967, p. 371.
4. Schalla, R. L.; McDonald, G. E., Ind. and Eng. Chem. 1953, 45, 1497-1500.
5. Thomas, A., Combustion and Flame 1962, 6, 46-62.
6. Glassman, I., "Phenamenological Models of Soot Processes in Combustion" Rept. 1450, Dept. of Mech. and Aerospace Engineering, Princeton Univ., 1979.
7. Cole, J. A., M. S. Thesis, Massachusetts Institute of Technology, Cambridge, Mass., 1982.

8. Cole, J. A.; Bittner, J. D.; Longwell, J. P.; Howard, J. B., Combustion and Flame, in press.
9. Bittner, J. D.; Cole, J. A.; Longwell, J. P.; Howard, J. B., to be published.
10. Bittner, J. D., Sc.D. Thesis, Massachusetts Institute of Technology, Cambridge, Mass., 1981.
11. Bittner, J. D.; Howard, J. B., 19th Symp. (Intl.) on Combustion, The Combustion Institute, 1983, p. 211.
12. "JANAF Thermochemical Tables" NSRDS-NBS37, 1971, 2nd ed.
13. Atkinson, R.; Darnall, K. R.; Lloyd, A. C.; Winer, A. M.; Pitts, J. N., Jr., Adv. Photochem. 1979, 11, 375-488.
14. Čvetanovic, R. J., Adv. Photochem. 1963, 6, 115-82.
15. Burke, E. J.; Brezinsky, K.; Glassman, I., "Preliminary High Temperature Studies of 1,3-Butadiene Oxidation" presented at the Eastern States Section/The Combustion Institute Meeting, Atlantic City, N.J., December 1982.
16. Tully, F. P.; Ravishankara, A. R.; Thompson, R. L.; Nicovich, J. M.; Shah, R. C.; Kreutter, N. M.; Wine, P. H., J. Phys. Chem. 1981, 85, 2262-9.
17. Nicovich, J. M.; Gump, C. A.; Ravishankara, A. R., J. Phys. Chem. 1982, 86, 1684-94.
18. Ellis, R. J.; Frey, H. M., J. Chem. Soc. 1966, A, 553-6.
19. Benson, S. W.; Shaw, R., Trans. Farad. Soc. 1967, 63, 985-92.
20. Cole, J. A., unpublished data.
21. James, D. G. L.; Suart, R. D., Trans. Farad. Soc. 1968, 64, 2735-51.
22. Benson, S. W.; Haugen, G. R., J. Phys. Chem. 1967, 71, 1735-46.
23. McMillen, D. F.; Golden, D. M., Ann. Rev. Phys. Chem. 1982, 33, 493-552.
24. Frey, H. M.; Walsh, R., Chem. Rev. 1969, 69, 103-24.
25. D'Alessio, A.; DiLorenzo, A.; Sarofim, A. F.; Beretta, F.; Masi, S.; Venitozzi, C., 15th Symp. (Intl.) on Combustion, The Combustion Institute, 1975, p. 1427.
26. D'Alessio, A.; DiLorenzo, A.; Borghese, A.; Beretta, F.; Masi, S., 16th Symp. (Intl.) on Combustion, The Combustion Institute, 1976, p. 695.
27. Crittenden, B. D.; Long, R., Combustion and Flame 1973, 20, 359-68.
28. Crittenden, B. D., Ph.D. Thesis, University of Birmingham, England, 1972.

RECEIVED November 30, 1983

Quantitative Chemical Mechanism for Heterogeneous Growth of Soot Particles in Premixed Flames

STEPHEN J. HARRIS

Physical Chemistry Department, General Motors Research Laboratories, Warren, MI 48090

In this article we present the first quantitative chemical mechanism for the heterogeneous growth of soot particles in premixed flames. We have found that the increased surface growth rate in sootier (richer) flames is due primarily to an increase in the surface area available for growth; the concentration of the gas phase growth species is similar from flame to flame. Growth decreases as the soot ages in the flame, but this is due to a decrease in the reactivity of the soot and not to a depletion of growth species. Acetylene supplies nearly all of the mass for soot growth, and our data suggest that soot growth can be understood in terms of a first order decomposition reaction of acetylene on the soot surface.

Soot formation in premixed flames may be divided into particle inception ("nucleation") and growth stages (1). In the nucleation stage tiny (1-2 nm) soot particles are created, while during the growth stage the soot particles coalesce and also accrete hydrocarbon molecules ("growth species") from the burned gases. These growth species react chemically with and become incorporated into the soot particles in a heterogeneous process known as surface growth, and they account for nearly all the final soot mass. Great efforts have been made towards understanding soot nucleation, and a number of mechanisms have been proposed (1); however, no comparable effort has been made towards understanding surface growth. In this article we propose the first quantitative chemical mechanism of soot particle surface growth in premixed flames. A more detailed account of this work will be published elsewhere (2).

0097-6156/84/0249-0023$06.00/0

Experimental

Premixed flames of C_2H_4 and a 79-21 mixture of Ar and O_2 were
stabilized on a water-cooled porous plug burner with a nitrogen
shroud. Because the flames were nearly 1-dimensional, measurements made as a function of height above the burner could be converted into measurements made as a function of time from knowledge of the hot gas velocity. Soot was detected using standard
Rayleigh scattering and extinction techniques (3,4) using an
argon laser, allowing soot number density, mean particle diameter,
and volume fraction (cm^3-soot/ cm^3-flame) to be determined. All
measurements were made far downstream from the soot nucleation
zone (The smallest particles detected had diameters of about 8
nm.) in a region of the flame where radical concentrations had
dropped to their equilibrium values (5) and where soot growth was
the main chemical process occurring. Temperatures, determined
from the brightness and emissivity of the soot, varied mainly
between 1650 and 1750 K. Samples of burned gases were collected
with a water-cooled quartz microprobe and batch analyzed for CO,
CO_2, C_2H_2, and CH_4 with IR and mass spectrometry. In the region
of the flame examined, Ar, CO, CO_2, H_2, and H_2O made up about 97%
of the gas phase material; C_2H_2, CH_4, and trace amounts of other
hydrocarbons made up the rest. Soot and condensable hydrocarbons
("volatiles") were collected on a water-cooled plate and subjected to thermogravimetric, elemental, and B.E.T. surface area
analysis (B.E.T. analysis determines the surface area of finely
dispersed particles by measuring the amount of N_2 adsorbed on the
particles at 77 K.)

Results

Two dynamical processes occur in the burned gas region of premixed
flames (4). First, small spherical soot particles collide and
coalesce into large spherical particles, a process which reduces
the total amount of surface area without changing the total mass.
The second process is surface growth, which accounts for an increase in the total soot mass and adds surface area. The net
effect is that while the number density falls and the mean diameter rises, the total soot surface area per cm^3 of flame, measured optically, changes very little as the soot ages. These
effects are displayed in Figures 1 and 2. However, Figure 2
shows that there is a steep increase in the soot surface area as
the flames become richer. Curves in the upper part of Figure 3
show soot volume fractions for five flames with different C/O
(carbon/oxygen) ratios as a function of time (soot age). All of
the increase in mass in each flame is attributed to surface
growth.
     To obtain surface growth rates we differentiate the volume
fraction curves, converting volume to mass using a density of 1.8
g/cm^3. Although all the curves appear to have similar slopes

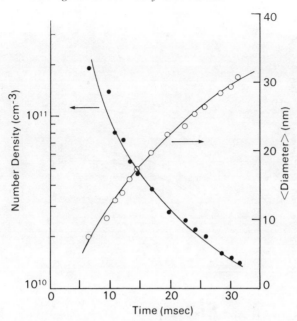

Figure 1. Mean diameter and soot particle density as a function of time (height above the burner) for a typical flame.

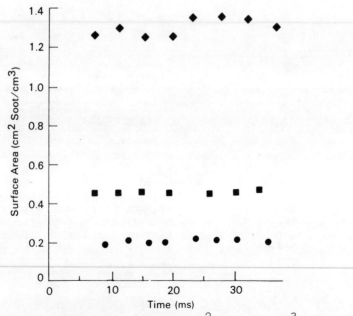

Figure 2. Total soot surface area ($\pi < d^2 > N$) per cm^3 of flame. The symbols refer to the stoichiometry: ●, -C/O = 0.76; ■, -0.82; and ◆, -0.94. (Reproduced with permission from Ref. 2. Copyright 1983, Gordon and Breach.)

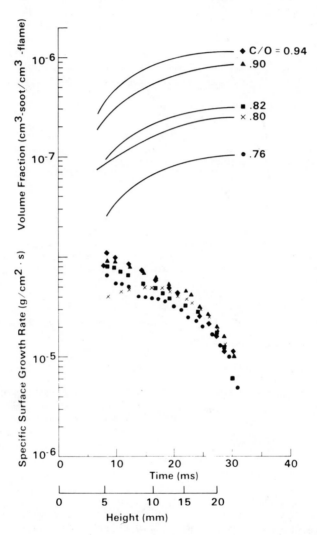

Figure 3. Top, soot volume fractions for flames with the indicated C/O ratios; and bottom, surface growth rates per cm^2 of soot surface. (Reproduced with permission from Ref. 2. Copyright 1983, Gordon and Breach.)

when plotted on semi-log paper, the derivatives are actually
quite different and increase very substantially with the C/O
ratio. However, of more fundamental interest than the total
growth rate (g/s) is the <u>specific</u> surface growth rate (g/cm^2-s),
which takes into account the greater surface area available for
growth in the richer (sootier) flames (Figure 2). Specific
surface growth rates, plotted in the lower half of Figure 3, vary
relatively weakly with C/O ratio compared to the total surface
growth rates.

Discussion

In order to interpret the surface growth results quantitatively,
we assume that surface growth is described by

$$dM/dt = S \ \Sigma \ k_i [g_i]^{j_i} \tag{1}$$

where M is the soot mass concentration, S is the total soot surface area per cm^3 of flame, $[g_i]$ is the mole fraction of the i'th
growth species, k_i is the rate constant for the surface reaction
which converts g_i into soot, and j_i is the reaction order. The
fact that the specific surface growth rates (1/S) dM/dt are similar in flames which have very different total growth rates and
which produce very different amounts of soot allows us to place
several significant constraints on the growth process and on the
surface growth species. (i) Surface growth is much faster in
richer flames primarily because there is more surface area available for growth; thus, growth species concentrations are similar
in the different flames. (ii) Growth rates fall steeply as the
soot ages, but this fall is not due to depletion of the growth
species. The reason is that if the $[g_i]$ are roughly independent
of the stoichiometry, and if they are high enough to supply all
of the mass increase in the richest flame, then there must be an
excess of them in all the leaner flames, where the increase in
soot mass is much less. Since the specific growth rate in the
richest flame is almost the same as in the leaner flames, there
is also no depletion of growth species in the richest flame.
(iii) From the total amount of soot mass picked up in the richest
flame, we calculate that either the growth species mole fractions
total more than 0.1/W, where W is the (mean) molecular weight of
the growth species, or that they are in <u>rapid</u> (compared to reaction with soot) equilibrium with a reservoir of other species
whose mole fractions total more than 0.1/W. Among the hydrocarbons in the burned gases, only methane and acetylene have mole
fractions that high. (iv) The youngest soot particles that we
detect have diameters of about 8 nm and molecular weights well
above 100,000 but contain only of the order of 10 nuclei (<u>6,7</u>).
Thus they are composed almost entirely of growth species. Since
there is no obvious way by which the young growing particles can
lose pure carbon, the C/H (carbon/hydrogen) ratio of the growth

species cannot be higher than that of the young soot, about 2 to 2.5 ($\underline{3}$). (As the soot ages its C/H ratio increases to about 8 ($\underline{3}$).)

To identify the growth species, we consider which hydrocarbons in the post flame gases satisfy the above constraints. The volatile material may easily be eliminated. For example, its concentration is about 100 times higher in benzene flames than in flames of aliphatic fuels ($\underline{8}$), but the growth rates are nearly identical ($\underline{9}$). Also, there is always much less volatile material than soot in the flame ($\underline{3}$), so there cannot be enough of it (constraints (ii) and (iii)) to account for the large increase in soot mass with age.

Polyacetylenes (C_4H_2 to $C_{10}H_2$ have been observed) are equilibrated with acetylene in flames ($\underline{10}$). Bonne et al. ($\underline{10}$) and Homann and Wagner ($\underline{8}$) argued that the largest of these are responsible for most of the surface growth, but their arguments were only correlations and were therefore not convincing. In fact, the evidence shows that they are not important. For example, except for diacetylene (C_4H_2), all of the polyacetylenes violate constraint (iv). Furthermore, in order for the higher polyacetylenes to be as important as C_4H_2, their k_i would have to be substantially higher since their concentrations are so much lower ($\underline{10}$). However, Tesner has found ($\underline{11}$) that in any homologous series of hydrocarbons an increase in molecular weight leads to only a "tiny" increase in the surface growth rates. Finally, deep in the post flame gases far beyond the nucleation zone, oxygen has been consumed and C_4H_2 will be formed from C_2H_2 by the pyrolysis mechanism modelled by Tanzawa and Gardiner ($\underline{12}$). For the richer flames the model gives a C_4H_2 formation rate only 20% of the peak measured rate of soot formation. This result means that diacetylene cannot be responsible for surface growth, since the equilibrium reactions connecting C_2H_2 and C_4H_2 would be too slow to prevent diacetylene from being depleted, violating constraints (ii) and (iii). Even if the model were not accurate enough to eliminate diacetylene on this basis alone, it also makes an important qualitative prediction. If diacetylene is importantly involved in soot growth, then destruction of diacetylene on the soot surface will substantially perturb the value of the equilibrium relationship between acetylene and diacetylene, $R = [C_4H_2][H_2]/[C_2H_2]^2$. However, values of R measured in heavily sooting ($\underline{10}$) and in nonsooting ($\underline{13}$) flames were essentially identical, implying that surface growth is in fact a negligible sink for C_4H_2.

Only C_2H_2 and CH_4 are present at high enough concentration to satisfy constraint (iii). However, given methane's concentration and the fact that methane and acetylene are not equilibrated in the flames ($\underline{13}$), CH_4 could not provide a major fraction of the material for soot growth without being substantially depleted, violating constraint (ii) and contradicting our concentration measurements. Thus, methane is not important for surface growth.

(Similarly, vinyl acetylene could not be important since it too is not equilibrated with acetylene (13).) Therefore, whether or not C_2H_2 itself actually decomposes on the soot surface, conservation of mass requires that practically all of the soot mass is provided, directly or indirectly, by the acetylene in the burned gases.

It remains to determine whether surface growth can be fully accounted for by direct decomposition of C_2H_2. Bonne et al. (10) argued that C_2H_2 could not be important since growth ceased at long times while the C_2H_2 concentration remained high. However, growth may cease even in the presence of the growth species if the soot becomes unreactive as it ages. The facts that as it ages soot loses its radical character (10) and that its chemical composition becomes more graphitic suggest that its reactivity does indeed fall. (Since the temperatures vary with stoichiometry by as much as 100 K, the substantial fall in the rate constant with time is not due primarily to the decrease in temperature with height above the burner, which is also about 100 K.) If we assume that acetylene is the principal growth species, then a plot of ln(specific growth rate) vs $ln[C_2H_2]$ gives a reaction order of j=0.9±0.7 (2 sigma); that is, the reaction is first order. The apparent first order rate constant k may then be determined, and it is plotted in Figure 4. The spread in the data which appears in the lower part of Figure 3 is largely eliminated.

Independent data on surface growth have been provided by Arefeva et al. (14). At 1700 K they measured a rate constant of 10^{-4} g/cm^2-s-atm for surface growth on pure carbon from C_2H_2, which is similar to the rate constant that we calculate for old (high C/H ratio) soot in our flame. The agreement supports the conclusion that surface growth is contributed primarily by acetylene.

Finally, we note that since the specific growth rates are similar in flames which produce very different amounts of soot, the processes controlling the ultimate soot loading must occur prior to the growth stage; that is, during the nucleation stage. Thus, even though surface growth is responsible for practically all of the mass of mature soot particles, it appears that richer flames produce more soot because they have a higher nucleation rate and therefore more surface area from the beginning of the growth stage.

Summary and Conclusions

(1) We have found that the increased surface growth in richer flames is accounted for primarily by the increased surface area available for growth in these flames, and not by a higher concentration of growth species. Thus, richer flames are sootier because they have a higher nucleation rate.

(2) Depletion of growth species does not occur in our flames. Therefore, the final size reached by the soot particles, when sur-

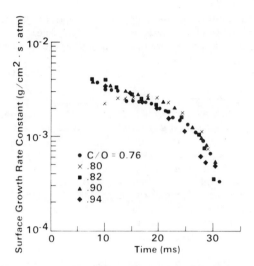

Figure 4. Apparent first-order rate constant for the reaction converting acetylene to soot. (Reproduced with permission from Ref. 2. Copyright 1983, Gordon and Breach.)

face growth has virtually ceased, is not determined by depletion but rather by a decrease in the reactivity of the soot.

(3) We have shown that the acetylene in the burned gases is the source for most of the mass of mature soot particles. Furthermore, we have found that the measured growth rates are consistent with the assumption that a first order decomposition reaction of acetylene with the soot surface is responsible for most of the soot growth, although diacetylene may also play some role.

Literature Cited

1.   Gaydon, A. G.; Wolfhard, H. G. "Flames" 1979, Chapman and Hall, 4th edi., London.
2.   Harris, S. J.; Weiner, A. M. Combustion Science and Technology 1983, 31, 155. Ibid, 32, 267.
3.   D'Alessio, A.; DiLorenzo, A.; Sarofim, A. F.; Beretta, F.; Masi, S.; Venitozzi, C., 15th Symposium (International) on Combustion, The Combustion Institute, 1975, 1427.
4.   Haynes, B.; Wagner, H. Gg. Energy and Combustion Science 1981, 7, 229.
5.   Millikan, R. C. J. Phys. Chem. 1962, 66, 794.
6.   Howard, J. B.; Wersborg, B. L.; Williams, C. G. in Faraday Symposium No. 7, Fogs and Smokes, Faraday Division, Chemical Society, Longon, 1973, 109.
7.   Smith, G. W. Combustion and Flame 1982, 48, 265.

8. Homann, K. H.; Wagner, H. Gg., 11th Symposium (International)
    on Combustion, The Combustion Institute, 1967, 371.
9. Haynes, B. S.; Jander, H.; Wagner, H. Gg. Ber. Bunsenges.
    Phys. Chem. 1980, 84, 585.
10. Bonne, U.; Homann, K. H.; Wagner, H. Gg., 10th Symposium
    (International) on Combustion, The Combustion Institute,
    1965, 503.
11. Tesner, P. A. Comb. Expl. Shockwaves 1979, 15, 111.
12. Tanzawa, T.; Gardiner, W. C. J. Phys. Chem. 1980, 84, 236.
13. Bittner, J. P. Ph.D. Thesis, Massachusetts Institute of
    Technology, Department of Chemical Engineering, 1980.
14. Arefeva, E. F.; Rafalkes, I. S.; Tesner, P. A. Khimiya
    Tverdogo Topliva (Solid Fuel Chemistry) 1977, 11, 113.

RECEIVED October 26, 1983

Ion Concentrations in Premixed Acetylene–Oxygen Flames near the Soot Threshold

D. G. KEIL, R. J. GILL, D. B. OLSON, and H. F. CALCOTE

AeroChem Research Laboratories, Inc., Princeton, NJ 08542

Total ion concentration profiles were measured through the flame front of laminar, premixed, low pressure acetylene–oxygen flames at equivalence ratios, ϕ, from 1.5 to 4.0. Peak total ion concentrations decrease with increasing ϕ up to and beyond the soot threshold ($\phi = 2.4$). At close to the soot threshold, a second peak appears in the ion profiles. This second peak contains some of the same ions, e.g., $C_{13}H_9^+$, as the first peak which is ascribed to chemi-ionization. The second peak reaches a minimum, at $\phi = 2.9$, well on the fuel rich side of the soot threshold, and then rises again with increase in equivalence ratio. In a sooting, $\phi = 3.0$, flame the concentrations measured are in reasonable agreement with measurements by others using molecular beam sampling techniques. The results refute the arguments against an ionic mechanism based on reported observations that the total ion concentration increases at the soot threshold. The measured ion concentrations are consistent with an ionic mechanism of soot nucleation.

Ions are observed in all premixed hydrocarbon flames. In near stoichiometric flames, the major ions are small (molecular weight < 100 amu); they result from the primary chemi-ionization reaction $CH + O = CHO^+$ and subsequent ion-molecule reactions. Recent flame ion sampling mass spectrometric studies (1) of premixed C_2H_2/O_2 flames have shown that as the flames are enriched to near or beyond soot threshold, the lighter mass ions (< 300 amu) decrease in concentration relative to the growth of larger ions (> 300 amu). This has been cited as evidence that rapid ion-molecule reactions convert small hydrocarbon "precursor" ions (e.g., $C_3H_3^+$, $C_5H_3^+$, and $C_7H_5^+$) to larger and larger ions which eventually become soot particles (2,3). The intermediate size ions with masses between about 100-500 amu correspond to polycyclic aromatic hydro-

0097–6156/84/0249–0033$06.00/0

carbon (PCAH) ions. The higher mass ions are less well identified, being generally grouped in mass ranges, but are assumed to
involve continued growth to soot particles. In studies of sooting
C_2H_2/O_2 flames, Homann (4) and Homann and Stroefer (5) have obtained low resolution ion mass data to about 10^5 amu, covering the
range from purely molecular ions to charged particles observable
by electron microscopy (equivalent to spheres with diameters of
about 6 nm). Although they found the mean ion mass increased
throughout the flame, they did not associate these ions with the
low mass PCAH ions since the total ionization increased sharply at
the height in the flame for the onset of sooting. Due to the
fairly low ion masses observed in this region (i.e., high ionization potentials), thermal ionization was also ruled out. Several
possible mechanisms for the production of such large molecular
ions have recently been considered by Calcote (6) who concluded
that continued growth of smaller chemi-ions by reactions with
major flame neutral species such as C_2H_2 or C_4H_2 is the most likely mechanism for the production of large molecular ions in flames.
On the other hand, Goodings, Tanner, and Bohme (7) and Michaud,
Delfau, and Barassin (8) favor buildup of large neutral species by
free radical reactions followed by an ion-molecule reaction of a
small ion with the neutral species to produce a large ion. Homann
and Stroefer favor thermal ionization of the large neutral species
by energy accumulation as the species grows (5).

Delfau et al. (9) measured the total ion concentration profiles in a series of low pressure $C_2H_2/O_2/N_2$ premixed flames from
nonsooting to heavily sooting using Langmuir probes and a molecular beam Faraday cage technique. On the fuel lean side of the
soot threshold, the peak ion concentrations decreased with increasing equivalence ratio whereas on the fuel rich side, the peak
ion concentrations increased with equivalence ratio. Since the
minimum peak ion concentration seemed to occur near the critical
equivalence ratio for soot formation, Delfau et al. and then
Haynes and Wagner (10) concluded that the ions in the sooting
flames are the result of soot formation and not the cause. Close
examination of the results of Delfau et al. (9), however, dis-
closes that the ion concentration reaches a minimum value somewhat
richer than the critical equivalence ratio for soot formation, ϕ_c;
compare $\phi_c = 2.2$ with ϕ for minimum ion concentration of 2.3.
Additionally, there have been questions about the value of the
total ion concentration in near sooting and sooting flames, compared to the ultimate soot particle concentrations (10-12).

These conflicting interpretations of the ionization behavior
near soot thresholds--both with variations of the equivalence
ratio and the peak locations in the flame-- and the questions
about the total ion concentrations prompted us to measure absolute
ion concentration profiles in the well studied 2.7 kPa (50 cm s^{-1}
unburned gas velocity) premixed C_2H_2/O_2 flame in an attempt to
better define the threshold behavior and to determine the absolute
ion concentrations.

Experimental Apparatus and Procedures

Low Pressure Flame Apparatus. All experiments were performed on
flat, premixed low pressure, 2.7 kPa (20 Torr) C_2H_2/O_2 flames stabilized on a water cooled burner. This 8.6 cm diam burner was
constructed of approximately 900, 0.12 cm i.d. stainless steel
tubes microbrazed into two stainless steel perforated plates to
form a water jacket around the tubes. Gases were metered using
calibrated critical flow orifices. The burner was installed in a
low pressure vessel pumped by a 140 L s^{-1} (300 CFM) mechanical
vacuum pump. Unburned gas velocities (298 K and 2.7 kPa) in all
cases were 50 cm s^{-1}. Equivalence ratios from ϕ = 1.5 to 4.0 were
studied with most emphasis on a sooting ϕ = 3.0 flame. Visible
soot emission became apparent at a soot threshold of ϕ = 2.4 to
2.5.

Langmuir Probe. A water-cooled electrostatic probe was constructed from several telescoping brass tubes around a 0.124 cm i.d.
stainless steel tube containing a Pt/10% Rh probe wire insulated
by fine quartz and Teflon sleeves. The probe wire, usually 0.025
cm diam, protruded 0.1 to 1.0 cm from the end of the tubes exposed
to the flame gases. This probe assembly is referred to as the
"fixed probe." A similar assembly was also used in which the Pt/
Rh wire could be quickly extended into the flame from an enclosing
housing by fixed amounts (0 to 1 cm) or withdrawn into the cooled
housing by externally moving a 0.127 cm diam Pt/Rh wire buttwelded
to the fine probe wire. A vacuum tight sliding O-ring seal was
used around the larger wire. This "pulsed" probe apparatus mini-
mized soot deposition on the probe wire by inserting the wire in
the flame just long enough to take a measurement. This probe was
used at a distance of 0.5 cm or farther from the burner, where
soot deposition on the probe was a problem.
   In both configurations, the probe was biased relative to the
grounded burner with a variable d.c. power supply. Probe currents
were measured with an electrically floating Keithley Model 602
electrometer. Probe resistance to ground was typically 10^{13} ohms,
measured before and after each experiment.
   Full current-voltage profiles in low pressure C_2H_2/O_2 flames
with ϕ = 1.5 to 3.5 were measured using the "fixed" probe to determine the probe current at plasma potential, V_p, by extrapolation.
The "pulsed" probe was used only at constant voltages (-20 and
-40 V) in order to minimize exposure times ($\approx$1-2 s) to the flame
and was used mainly under sooting conditions, $2.4 \gtrsim \phi \leq 4.0$.

Probe Theory. Experimental probe currents at the plasma potential,
where diffusive transport of ions to the probe dominates convective transport, were converted to ion concentrations using the
equation derived by Calcote (13):

$$n_+ = \frac{(I_+)_{pl}}{2e\ \pi r_p \ell_p} \left(\frac{2\pi\ m_+}{kT}\right)^{1/2} \left[1 + \frac{1.5\ r_p \ell_p}{4\lambda_+\ B}\ \ln\left(\frac{X + B}{X - B}\right)\right] \quad (1)$$

Here n_+ is the ion concentration (cm^{-3}) of ions with mass m_+ and mean free path λ_+. $(I_+)_{pl}$ is the current at plasma potential collected by a cylindrical probe wire of radius r_p and length ℓ_p. The symbols e, k, and T represent the electron charge, Boltzmann's constant, and the temperature (K), respectively. The parameters X and B are defined: $X = \ell_p + 2\lambda_+$ and $B = (X^2 - 4(r_p + \lambda_+)^2)^{1/2}$.

Probe currents measured at high negative biases were converted to ion concentrations using the Clements and Smy thick sheath theory (14) for ion collection in a flowing plasma under conditions where convective rather than diffusive transport of ions to the probe region dominates:

$$I_+ = \frac{2\ell_p(\pi\mu_+\ \epsilon_0)^{1/3}\ (n_+\ e\ v_f V_p)^{2/3}}{[\ln(I_+/2n_+ e\ v_f\ r_p \ell_p)]^{2/3}} \quad (2)$$

where μ_+ is the ion mobility, ϵ_0 is the permittivity of space, v_f is the local flow velocity, and I_+ is the probe current collected at negative V_p volts relative to plasma potential.

Since the assumed mode of ion transport to the probe wire differs in the derivation of Equations 1 and 2, the most appropriate type of measurement depends on the local flame conditions, the degree of ionization, the mass distribution of ions, etc. Estimates of this information were used along with certain probe theory criteria to select the most accurate method for each individual measurement (15).

The application of Equations 1 and 2 requires information on flame temperatures, flow velocities, and ion mobilities/mean free paths throughout the flame, as well as the probe wire dimensions. This information is used both to select the type of measurement to make (e.g., probe currents at plasma potential or at large negative potentials) and to calculate the ion concentration, n_+. Details of probe theory selection as well as the determination of the necessary local flame and ion transport properties are given elsewhere. The approach, however, is summarized here.

Flame temperature profiles against distance from the burner are based on sodium line reversal measurements of Bonne and Wagner (16) in similar 2.7 kPa, 50 cm s^{-1} C_2H_2/O_2 flames. In flames not measured by Bonne and Wagner, the profiles were scaled with calculated adiabatic flame temperatures. An uncertainty of 100 K in the temperature results in a minor (< 10%) error in the calculated ion concentrations. Flow velocities are determined from the local temperature and equilibrium composition (mole change).

Ion transport properties are estimated on the basis of the ion masses in the flame. The method used to estimate these numbers is not straightforward since an extremely wide range of ion

masses is observed in sooting flames. We have considered ion
masses observed in our flame ion mass spectrometric studies of low
mass ions (< 300 amu) and in the heavy flame ion (100 < mass < 10^5
amu) studies by Homann and coworkers (4,5) using their variable
high-pass ion mass filter in sooting flames (ϕ = 2.8–3.25).

Our mass spectroscopic studies have shown that in nonsooting
flames the dominant ions are either H_3O^+ (19 amu) or $C_3H_3^+$ (39
amu) with $C_3H_3^+$ dominating in the richer of the nonsooting flames.
In sooting flames ($\phi \geqq 2.4$) $C_3H_3^+$ also dominates within about 1 cm
from the burner. Beyond this region, the ion masses rapidly increase so we must depend on the studies of Homann et al.

The recent ion mass results of Homann and Stroefer (5) were
scaled to Homann's more complete results (4) (heavy ion mass distributions determined at more distances from the burner in his ϕ =
2.9, 2.7 kPa flame). The resultant profile of median ion mass
versus distance from the burner was shifted 0.2 cm closer to the
burner (based on Homann and Stroefer's experiments on the effect
of flow velocity on ion profiles) to correct for the small velocity difference between their unburned gas velocity (≈ 43 cm s^{-1})
for which they report data and our unburned gas velocity (50 cm
s^{-1}). This shift is consistent with their observed shift of the
peak ion signal location with velocity. Since Homann and Stroefer
found little difference in the 2 cm mass distribution between ϕ =
2.8 and ϕ = 3.5, the results should apply equally to a ϕ = 3.0
flame. "Smooth" extrapolations of the ion masses were made beyond
4 cm and to mass 39 at 0.9 cm.

This procedure resulted in the median ion mass profile shown
in Figure 1 as a smooth curve. Also shown are three points of
median charged particle masses from the electron–microscope study
of Adams (17). A particle density of 1.5 g cm^{-3} was used to convert Adams' reported particle diameter to masses. There is good
agreement within a factor of 2 between the charged particle masses
and the median ion mass profile. The greatest uncertainties in
the median ion mass are seen to occur at distances between 1 and 2
cm from the burner where the mass increases by more than two
orders of magnitude.

In the reduction of our Langmuir probe data we assumed this
ion mass profile in all sooting flames with $\phi \geqq 2.75$. In the
slightly sooting ϕ = 2.5 flame, we used the intermediate profile
shown as a dashed line in Figure 1, based on a somewhat arbitrary
geometric mean between the ion masses in a nonsooting ϕ = 2.0
flame (m$_+$ = 39 amu) and those in a heavily sooting flame. The
uncertainties introduced by these approximations will be discussed
with presentation of the results of this study.

Our mass spectrometric studies have shown that major ions <
500 amu are consistent in mass with polycyclic aromatic hydrocarbon (PCAH) ions. Experimental ion mobilities of a wide mass range
of PCAH ions (18,19) were extrapolated to very high masses in order to estimate the mobilities of the heavy median mass ions (Figure 1). This extrapolation parallels a Langevin-type equation,

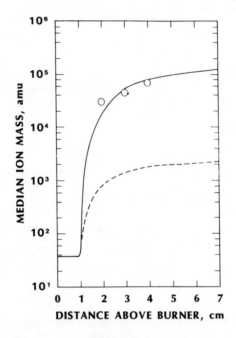

Figure 1.   Estimated median ion masses based on Homann and coworkers heavy ion masses (see text) in sooting flames.   Solid
line used to convert Langmuir probe currents to ion concentrations
in flames with $\phi > 2.5$; dashed line is geometric mean of solid
line and mass 39 and was used in $\phi = 2.5$ flame.   o, charged particle mass ($\underline{17}$); and density = 1.5 g cm^{-3}.

giving us confidence in the procedure. Corrections of the resultant ion mobilities were made for flame temperature, pressure, and for composition using Blanc's law.

Results and Discussion

Ion concentrations were derived from the experimental probe data based on the mass distributions in Figure 1 in sooting flames ($\phi \geq$ 2.5) except in regions close to the burner (0 to 1 cm) where 39 amu ($C_3H_3^+$) was used for the mass of the flame ions. In the leaner flames ($\phi < 2.4$), mass 39 was used in the calculations with the exception of the $\phi = 1.5$ flame in which our flame ion sampling mass spectral studies indicate that beyond about 1.2 cm, the dominant ion is H_3O^+ (mass = 19 amu). The resultant profiles are shown in Figure 2 where the dashed lines are based on Calcote's theory and the solid lines are based on the thick sheath theory of Clements and Smy. The error bar at 5 cm on the $\phi = 3.0$ profile is representative of the reproducibility of the absolute concentrations determined at different times. The reported profiles represent point by point averages. The peak ion concentrations in the sooting flames were generally more reproducible than were the concentrations farther away from the burner. The transitions between dashed profiles and solid profiles in sooting flames were subject to some smoothing, being based on separate experiments, but the overlap was within the uncertainties. The region near the burner in sooty flames gave the least reliable measurements, due in part to large temperature gradients, rapidly changing ion masses, and low ion concentrations.

The nonsooting flames ($\phi = 1.5$, 2.0) exhibit a single peak which decreases in magnitude and shifts downstream with increasing ϕ. At $\phi = 2.5$, near the threshold for sooting, ($\phi_c = 2.4-2.5$), a second peak considerably farther downstream begins to appear. This second peak occurs at a distance where the median ion mass is greatly increasing (Figure 1). The calculated relative ion concentrations at the first and second peak in this profile are dependent on the relative ion masses assumed in the peak regions through the ion mobilities used in Equation 2. For the heavy ions in the region of the second peak, we have found that the conversion factor from probe currents to ion concentrations is roughly proportional to $m_+^{0.18}$ due to the fairly weak dependence on μ_+. In general, this is a small mass effect, but it can be significant if the mass is changing rapidly; an order of magnitude increase in mass leads to a 50% greater calculated n_+. In richer, heavily sooting flames ($\phi = 3.0$, 3.5) the second peak becomes more accentuated while the first peak continues to decrease with increasing ϕ. In these flames, the ion probe currents show distinct double peaks. Thus, although the second peak in the $\phi = 2.5$ flame might be considered an artifact of the conversion of probe currents to ion concentrations, it clearly is not so in the richer flames. The magnitude of the second peak decreases as ϕ increases from 2.5 to

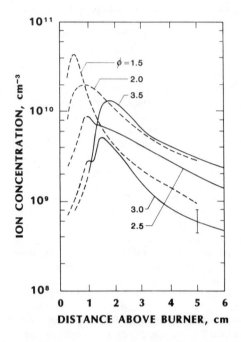

Figure 2. Total ion concentration profiles from Langmuir probe. Results in solid lines based on currents measured with large negative bias and Equation 2 (14); dashed lines based on currents at plasma potential and Equation 1 (13). Error bar at 5 cm for ϕ = 3.0 flame represents reproducibility of measurements at various times.

2.9. Further enrichment of the flame to ϕ = 3.5 reverses the trend with a large increase in the magnitude of the second peak.

The ion concentrations at the peaks are plotted as a function of equivalence ratio in Figure 3. Ignoring for the moment any differential between the first and second maxima, it is seen that peak ion concentration in a series of flames of increasing fuel richness beyond ϕ = 1.5 decreases well beyond (≈ 0.5 equivalence units) the critical threshold for soot formation, ϕ_c, indicated by the shaded area. This phenomenon is more marked than that observed by Delfau et al. (9) in their $C_2H_2/O_2/N_2$ flames where the lowest maximum occurred about 0.1 equivalence unit greater than ϕ_c. This observation refutes any argument against the ionic soot formation mechanism based on an increase in peak ionization at the soot threshold equivalence ratio.

As in the investigation by Delfau et al., the first maximum in the present study is associated with ions derived from normal flame chemi-ionization processes because, as the flame is made increasingly fuel rich, the position of the maximum varies continuously and its magnitude decreases with increasing ϕ even in the sooting flames as one would expect. The second maximum is explained by Delfau et al. (9) as due to "direct thermal ionization of large polynuclear aromatic molecules and carbon particles less than 2 nm diameter of low ionization potentials." This explanation has already been questioned by Howard and Prado (20). In the Delfau et al. paper and the referenced discussion, the second peak is identified with the "ionized particle" peak (a third peak) which will be discussed below. Examination of Figure 5 in the paper of Delfau et al. indicates that their first and second peaks correspond to our first and second peaks which both occur, as will be shown, before the peak in "charged particles" as identified by Wersborg, Yeung, and Howard (21).

Another difficulty with the Delfau et al. explanation is that the second peak contains many of the same small ions as the first peak; see e.g., Figure 4. We have identified several ions in the second peak including: $C_{13}H_9^+$, $C_{15}H_{11}^+$, $C_{18}H_{11}^+$, $C_{19}H_{11}^+$, and ions containing 21, 23, 25, 27, and 29 carbon atoms. These ions are far too small to be produced by thermal ionization and no mechanistic path has been identified for the production of small ions from large ions--or charged particles (6).

The locations of the two maxima are plotted as a function of the equivalence ratio in Figure 5. In the nonsooting flames, the (first) maxima tend to move away from the burner with increasing equivalence ratio, while the second maxima behave similarly in the sooting flames.

Figure 6 shows a comparison of the total ion concentration profile determined in the present study of the ϕ = 3.0 flame with relevant data from other studies. These include profiles of total ions, charged particles, and neutral species. The profile indicated by HS represents molecular beam Faraday cage measurements by Homann and Stroefer (5). The profile is actually an average of

CHEMISTRY OF COMBUSTION PROCESSES

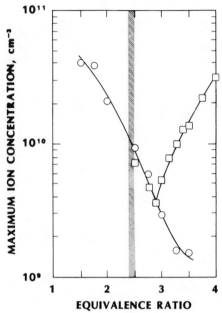

Figure 3. Maximum total ion concentrations as a function of equivalence ratio. Same flames as Figure 2. o, maxima closest to burner; and □, second maxima. Soot threshold is indicated by shading.

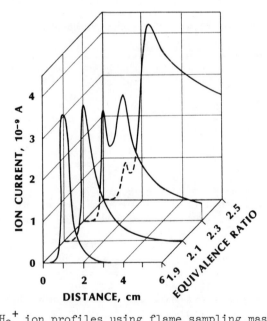

Figure 4. $C_{13}H_9^+$ ion profiles using flame sampling mass spectrometer technique in C_2H_2/O_2 flame (2.1 kPa, 50 cm s^{-1} unburned velocity). Critical threshold, $\phi_c \approx 2.5$.

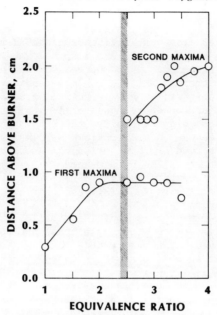

Figure 5. Locations of the two ion concentration maxima in Figure 3. Soot threshold indicated by shading.

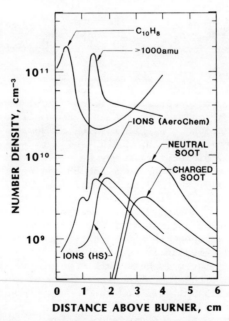

Figure 6. Comparison of present φ = 3.0 Langmuir probe total ion concentration profile with other results. AeroChem, Langmuir probe; HS, molecular beam Faraday cage (5) (see text); neutral soot and charged soot (21); $C_{10}H_8$ and > 1000 amu, neutral species (22).

their results for $\phi = 2.8$ and $\phi = 3.25$ flames, which are similar
in shape, the richer flame exhibiting somewhat higher ion concentrations. The resultant profile has also been shifted 0.2 cm,
vide supra, towards the burner to correct for the small velocity
difference between their flame and the other 50 cm s^{-1} flames
represented in the figure. There is overall good agreement between their total ion concentration profile and the Langmuir probe
result. HS, however, did not observe the first peak 1 cm from the
burner, probably because their technique was less sensitive to the
lower mass ions observed in this region of the flame. They may
not have looked closer to the burner because, as already discussed,
this is a difficult region in which to take measurements (see
Homann's comments following Ref. 22), and they had no reason to
expect another peak there.

Figure 6 shows two neutral species profiles from the molecular beam mass spectrometric study by Bittner and Howard (22). The
polycyclics $C_{10}H_8$ and, at about an order of magnitude lower concentration, $C_{14}H_8$ (not shown) peak early in the flame at about 0.5
and 0.7 cm, respectively, while large neutrals > 1000 amu (relative concentration scale) peak in the region of the second ion
maximum. Homann and Wagner (23) also observed a peak of large
neutral species they called "soot precursors" in the region where
the second ion maximum is observed. The general correspondence of
two maxima in both the ion concentration and the large neutral
species in sooting flames raises several questions. Are these
ions and neutrals associated with each other? If so, are the ions
the source of the neutrals, or vice versa? They certainly appear
to have a common genesis. Does the appearance of the second ion
maximum herald soot formation? These questions remain unanswered
at this time.

Also shown in Figure 6 are profiles for large (>1.5 nm) soot
and charged particles from molecular beam electron microscopy
studies of Howard and coworkers (21). As shown above, the masses
associated with the charged particles agree well with the median
mass ion profile shown in Figure 1 based on Homann and Stroefer's
work. The differences between the HS and AeroChem probe profiles
and the charged particle profile may be explained by arguing that
the two techniques used to produce the HS and AeroChem profiles do
not respond to heavy, charged soot particles, while the electron
microscope study used to produce the charged soot profile does not
detect small ions. However, these discrimination effects must be
greater than might be estimated if the three profiles are located
consistently with respect to the burner. At 2 cm, Homann and
Stroefer observed a median ion mass of roughly 10^4 amu corresponding to an "ion" diameter (spheres with density 1.5 g cm^{-3}) greater
than 2.5 nm, within the reported sensitivity of the electron microscope studies (>1.5 nm). If the three measurement techniques,
Langmuir probe, Faraday cage, and electron microscope, measure the
same species, why do the measurements differ by an order of magnitude at this distance? Even allowing an effective lower detecta-

bility limit of 2.5 nm with the electron microscope measurement,
the mass distribution from the Homann and Stroefer study at 2 cm
(50% of the ions > 2.5 nm) indicates that the total ion concentrations for the three measurements should be in much closer agreement. The differences are probably attributable to experimental
error and to the fact that three somewhat different systems were
studied.

If consistency is evoked at 3.5 cm from the burner by arbi-
trarily shifting the Faraday cage (5) and Langmuir probe profiles
in Figure 6 to higher concentrations relative to the Howard et al.
particle measurements (21), so that all three curves match at this
distance, then the peak ion concentrations in the shifted profiles
are 1.0 to 1.4 x 10^{10} cm^{-3}, compared to the peak neutral plus
charged particle concentration of 1.2 x 10^{10} cm^{-3}. The ion concentration is sufficiently large to provide support for the hypothesis
of ionic precursors for the soot concentration observed in the
flame because the peak ion concentration is comparable to the final
soot concentration. It should be noted, however, that ion soot
precursor questions are ultimately more strongly related to fluxes
and rates of production than to concentrations.

The corrected ion and charged particle profiles in Figure 6,
coupled with the information in Figure 1, could be looked at as a
smooth progression from small ions to large molecular ions to large
charged soot particles which produce neutral particles on recombination, consistent with an ionic mechanism of soot formation.

In summary, we have measured total ion concentration profiles
in a series of premixed, low pressure C_2H_2/O_2 flames (2.7 kPa, 50
cm s^{-1} unburned velocity). In rich sooting flames, the ion concentration is observed to maximize at two distances from the burner.
The maximum closest to the burner is attributed to normal chemi-
ionization while the second maximum has been shown to contain many
large ions as well as a number of the smaller ions observed in the
first maximum. With increasing fuel richness from nonsooting to
heavily sooting flames, the peak ion concentration in the first
maximum decreases monotonically. The ion concentration at the
second maximum, which begins to appear in flames near the critical
equivalence ratio for soot formation (ϕ_c = 2.4 to 2.5), also decreases with fuel enrichment well beyond ϕ_c. This differs from the
observations reported by Delfau et al. (9) for a series of increasingly richer low pressure $C_2H_2/O_2/N_2$ flames in which the ion
concentration in the second maximum is lowest at an equivalence
ratio just to the rich side (about 0.1 equivalence units) of ϕ_c.
The source of the second maximum (in ions and neutrals) is not
clear but it seems to be related to soot precursors.

The peak ion concentration in the well-studied ϕ = 3.0 flame
(2.7 kPa, 50 cm s^{-1}) has been established to be about 5 x 10^9 cm^{-3}
by two vastly different techniques in two laboratories. This concentration is comparable to the neutral and charged soot concentrations determined by a third technique in a third laboratory.
Differences in spatial ion and charged particle concentration pro-

files can be at least partly attributed to discrimination effects
in the various techniques used. The ion concentrations are sufficiently high to provide ionic precursors for the soot produced
later in this flame. We conclude that the ionic soot formation
mechanism remains viable.

Acknowledgments

This work was supported by the Air Force Office of Scientific Research (AFSC) under Contract F49620-81-C-0030. The United States
Government is authorized to reproduce and distribute reprints for
governmental purposes notwithstanding any copyright notation here-
on.

Literature Cited

1. Olson, D.B.; Calcote, H.F., in "Eighteenth Symposium (Inter-
    national) on Combustion"; The Combustion Institute: Pitts-
    burgh, 1981; p. 453.
2. Olson, D.B.; Calcote, H.F., in "Particulate Carbon: Formation
    During Combustion"; Siegla, D.C.; Smith, G.W., Eds.; Plenum:
    New York, 1981; p. 177.
3. Calcote, H.F. Combust. Flame 1981, 42, 215.
4. Homann, K.H. Ber. Bunsenges. Phys. Chem. 1978, 83, 738.
5. Homann, K.H.; Stroefer, E., in "Soot in Combustion Systems
    and Its Toxic Properties"; Lahaye, J.; Prado, G., Eds.;
    Plenum: New York, 1983; p. 217.
6. Calcote, H.F., in "Soot in Combustion Systems and Its Toxic
    Properties"; Lahaye, J.; Prado, G., Eds.; Plenum: New York,
    1983; p. 197.
7. Goodings, J.M.; Tanner, S.C.; Bohme, D.K. Can. J. Chem. 1982,
    60, 2766.
8. Michaud, P.; Delfau, J.L.; Barassin, A., in "Eighteenth Sym-
    posium (International) on Combustion"; The Combustion Insti-
    tute: Pittsburgh, 1981; p. 443.
9. Delfau, J.L.; Michaud, P.; Barassin, A. Combust. Sci. Techn.
    1979, 20, 165.
10. Haynes, B.S.; Wagner, H.Gg. Progr. Energy Combust. Sci. 1981,
    7, 229.
11. Olson, D.B.; Calcote, H.F., in "Eighteenth Symposium (Inter-
    national) on Combustion"; The Combustion Institute: Pitts-
    burgh, 1981; Comments by Wagner, H.Gg. and authors' reply,
    p. 463.
12. Olson, D.B.; Calcote, H.F., in "Particulate Carbon: Formation
    During Combustion"; Siegla, D.C.; Smith, G.W., Eds.; Plenum:
    New York, 1981; Comments by Ulrich, G.D. and Lahaye, J., pp.
    201, 203.
13. Calcote, H.F., in "Eighth Symposium (International) on Com-
    bustion"; Williams and Wilkins: Baltimore: 1962; p. 184.

14.   Clements, R.M.; Smy, P.R. J. Appl. Phys. 1969, 40, 4553.
15.   Keil, D.G.; Olson, D.B.; Gill, R.J.; Calcote, H.F., in prepa-
      ration for Combust. Flame, 1983.
16.   Bonne, U.; Wagner, H.Gg. Ber. Bunsenges. Phys. Chem. 1965,
      69, 35.
17.   Reported in: Prado, G.P.; Howard, J.B., in "Evaporation-
      Combustion of Fuels"; Zung, J.T., Ed.; American Chemical
      Society: Washington, DC, 1978; p. 153.
18.   Griffin, G.W.; Dzidic, I.; Carroll, D.I.; Stillwell; R.N.,
      Horning, E.C. Anal. Chem. 1973, 45, 1204.
19.   Hagen, D.F. Anal. Chem. 1979, 51, 870.
20.   Howard, J.B.; Prado, G.P., Comments and Delfau et al. re-
      plies, Combust. Sci. Techn. 1980, 22, 189.
21.   Wersborg, B.L.; Howard, J.B.; Williams, G.C. in "Fourteenth
      Symposium (International) on Combustion"; The Combustion In-
      stitute: Pittsburgh, 1973; p. 929.
22.   Bittner, J.D.; Howard, J.B.; in "Particulate Carbon: Forma-
      tion During Combustion"; Siegla, D.C.; Smith, G.W., Eds.:
      Plenum: New York, 1981; p. 109.
23.   Homann, K.H.; Wagner, H.Gg., in "Eleventh Symposium (Interna-
      tional) on Combustion"; The Combustion Institute: Pittsburgh,
      1967; p. 371.

RECEIVED December 21, 1983

Reactivities and Structures of Some Hydrocarbon Ions and Their Relationship to Soot Formation

JOHN R. EYLER

Department of Chemistry, University of Florida, Gainesville, FL 32611

The reactions of three ions, $C_3H_3^+$, $C_5H_5^+$, and $C_6H_5^+$, which have been sampled from fuel-rich and sooting flames, have been studied with a variety of flame neutrals using an ion cyclotron resonance (icr) mass spectrometer. Three different mass spectrometric techniques have been used to differentiate isomeric forms of the ions whose reaction rate coefficients were measured. These include icr reactivity differences, reactive collisions in a triple quadrupole mass spectrometer, and collision-induced dissociaton reactions in a reversed-geometry, double-focusing mass spectrometer. As a complement to experimental work, theoretical calculations have been carried out to predict the visible and ultraviolet absorption spectra of several isomeric forms of $C_3H_3^+$ and $C_5H_5^+$, as well as the relative stability of a number of isomeric forms of the latter ion. The structure determination and the ion/molecule reactivity studies both provide results consistent with and supportive of recently proposed ionic mechanisms for soot nucleation.

The existence of ions in flames has been known for many years, with Langmuir probes (1) and mass spectrometric sampling (2) used in a number of the early investigations. However, ion number densities were found to be much higher than could be explained by equilibrium thermal ionization at the temperatures occuring in flames. Calcote (3) first suggested the chemiionization mechanism

$$CH^* + 0 \rightarrow HCO^+ + e^- \qquad (1)$$

which is now generally accepted as the most important ion formation mechanism in many flames. In recent years a number of groups (4-7) have used improved mass spectrometric sampling techniques to obtain detailed profiles of ions in both fuel-lean and -rich flames, and to explain the ion chemistry involved in the formation and loss of the observed ions (8). Since most combustion processes have been shown to proceed via neutral and free-radical reactions, the importance of ions in combustion (as opposed to their existence) has not been established. One area of combustion chemistry where ions and ion/molecule reactions may play a critical role, and which has been emphasized in our studies to date, is that of soot nucleation in fuel-rich flames.

A recent review by Calcote (9) discusses many of the proposed mechanisms of soot nucleation, both neutral and ionic, and presents the case for an ion/molecule scheme, beginning with $C_3H_3^+$ and sequentially adding primarily acetylene and polyacetylene molecules in rapid condensation and condensation-elimination reactions which lead to polycyclic aromatic hydrocarbon ions of m/z 500 - 1000 amu. This particular ion/molecule soot nucleation model has been elucidated further in an article by Calcote and Olson (10) where a series of ion/molecule reactions were combined with acetylene oxidation reactions and a computational model developed. This model gave ion profiles reasonably similar to those actually observed in sooting flames, and also predicted concentrations of ions sufficiently high that they might be considered as soot nucleation sites. Aspects of an ionic mechanism for soot nucleation are discussed in more detail in another article in this volume, so no more will be said at this point about the general features of the scheme. It should be noted, however, that objections to it have been raised recently (11).

Our work in this area has been prompted by a number of weaknesses of the ion/molecule soot nucleation scheme of Calcote and Olson (10) (noted by the authors in their article). Very few experimental studies have been carried out to determine the rate coefficients for reactions of small hydrocarbon ions with various flame neutrals. Noting this lack of data, Calcote and Olson set all rate coefficients in their model to the (not unreasonable) value of 2×10^{-11} cm^3/s. As will be seen later in this article, a large number of isomeric structures are possible for even small hydrocarbon ions, and little, if any, reliable thermochemical data exists for any of them. Calcote and Olson, in their model, chose what seemed to be the most reasonable structure for many of the ions involved, and estimated heats of formation in the many cases where none could be found in the literature.

In an attempt to provide more accurate data on ion/molecule reactions, ion structures, and ion thermochemistry which may be relevant to soot formation, we have initiated a program to obtain such information in controlled laboratory studies. Ion/molecule reaction rate coefficients for the reactions of major (and initially low molecular weight) ionic species found in fuel-rich and sooting flames with a variety of flame neutrals have been and are continuing to be determined. We have used several mass spectrometric techniques, including ion cyclotron resonance (icr) mass spectrometry, triple quadrupole tandem mass spectrometry, and collision-induced dissociation in a reversed-geometry ZAB-2F mass spectrometer to probe the structures of small hydrocarbon ions. And we have complemented our experimental studies with theoretical calculations of the energies and spectra of certain ions whose structures are not well established. The remainder of this article will discuss our work on three ionic systems which are seen in early stages of the ion/molecule chemistry in fuel rich flames: $C_3H_3^+$, $C_5H_5^+$ and $C_6H_5^+$.

Experimental

The majority of the studies to be reported here have been carried out using ion cyclotron resonance mass spectrometry. This variation of mass spectrometry and its many applications have been discussed in detail in several review articles (12-14) and at least one book (15). Briefly, it uses combinations of static electric and magnetic fields to trap gaseous ions at low pressures for time periods of up to several seconds in duration. During these long trapping periods, the ions can be subjected to electromagnetic radiation, often from tunable lasers, to study their spectroscopic properties, or their reactivity with various neutral compounds present in the trapping region can be followed. With the pulsed icr method (16) determination of ion/molecule reaction rate coefficients for quasi-thermal ions at temperatures from ca. 300 K to several hundred K higher can be made in a straightforward manner (17-19). The structures of isomeric ions can also be probed with this technique since ions of different structure often exhibit different reactivities toward selected neutrals. This approach was first employed by Gross and co-workers (20) in a study of $C_6H_6^+$ isomers, and has been utilized more recently by Ausloos and Lias (21-22) for $C_7H_7^+$ and $C_4H_4^+$ ions.

For one study involving the differentiaton of $C_3H_3^+$ isomers, a Finnigan triple quadrupole mass spectrometer was employed. This type of mass spectrometric instrumentation has been described in several publicatons (23-24) and has been applied to a number of problems (25-26) in analytical chemistry. The first quadrupole is used to select an ion of interest, the second, in an rf-only mode, is used as a trap in which the ion undergoes (in our case reactive) collisions with a selected neutral gas, and the third quadrupole is used to analyze the results of those collisions. Work on $C_6H_5^+$ structure differentiation was carried out on a VG Analytical Instruments ZAB-2F reverse-geometry double-focusing mass spectrometer (27) at the Naval Research Laboratory in Washington, DC. Using the MIKES-CID (28) approach, a particular m/z ion is passed through the magnetic sector of the spectrometer and, with kinetic energy usually in the 4-8 keV range, collides with a gas in the second field free region. The products of this collision are energy analyzed by an electrostatic analyzer, with this information then related to the mass of the collision products.

Theoretical studies on ion structures and spectra were carried out using initially the Amdahl 450-V7 and more recently the IBM 3081D computer of the Northeast Regional Data Center at the University of Florida, and a Digital Equipment Corp. VAX 11/780 minicomputer in the Quantum Theory Project at UF.

Results

$C_3H_3^+$. $C_3H_3^+$ is the major ion sampled from a wide range of fuel-rich and sooting flames, and has been taken as the starting point for the ionic soot formation scheme proposed by Calcote and Olson (although there is stil some uncertainty as to its exact formation mechanism in certain flames) (10). Two isomeric structures are important in discussing its role in flame systems. The first is the cyclopropenylium isomer, I, which has been most often formed

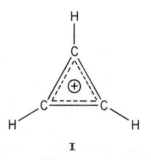

I

and studied in mass spectrometric and theoretical work to date. This is generally recognized as the most stable $C_3H_3^+$ isomer, with a theoretically calculated heat of formation of 253 kcal/mol (29), which is in quite good agreement with the 256 ± 2 kcal/mol determined (30) by experiment. $C_3H_3^+$ ions possessing this structure can be formed for study in a mass spectrometer by electron impact on a number of precursors, including allene (C_3H_4) and the various propargyl halides (C_3H_3X). A second and potentially more important $C_3H_3^+$ structure is that of the linear propargylium ion, II.

II

The heat of formation of this ion has been calculated ($\underline{29}$) to be 31-34 kcal/mol higher than that of the cyclopropenylium ion, again in fair agreement with the 25 kcal/mol difference found ($\underline{30}$) experimentally. Far less theoretical and experimental attention has been given this form of the $C_3H_3^+$ ion, although there was at least one suggestion ($\underline{31}$) that it might be important in flames. Mass spectrometric study of the propargylium isomer has become more straightforward with the report ($\underline{32}$) by Ausloos and Lias that significant fractions of the isomer can be produced by charge transfer reactions of small ions (Ar^+, Xe^+, CO^+, Ne^+, etc.) with propargyl chloride and bromide. Work in our laboratories has shown that even higher proportions of II relative to I can be obtained by either electron impact on or charge exchange with propargyl iodide (synthesized by a halide exchange reaction with propargyl bromide). Other isomeric $C_3H_3^+$ structures have been calculated ($\underline{29}$) to be 1 eV or higher in energy than II, and have not been seriously studied or discussed in connection with flame mechanisms.

Because of its stability, the cyclopropenylium ion, I, has been thought to be relatively unreactive toward simple hydrocarbon fuels. This has been confirmed in a study ($\underline{33}$) by workers from the National Bureau of Standards, who found that I was unreactive with ethylene, allene, and more importantly, acetylene and diacetylene. This isomeric form of the ion did show moderate reactivity toward some hydrocarbons such as iso-C_4H_8, trans-2-pentane, and 1,3-cyclo-C_6H_8. Ions with the structure II were quite reactive with many of the 26 neutral species studied. In particular, reaction of II with acetylene produced $C_5H_3^+$ and $C_5H_5^+$ ion populations with reactive and unreactive components. The reactive isomers of these ions combined with acetylene to form $C_7H_x^+$ ions, which probably possessed a stable, cyclic structure. It thus appears that if an ionic soot formation mechanism similar to that proposed by Calcote ($\underline{9}$) is important in combustion systems, the linear, propargylium ion, II, is the important reactive precursor.

A suggestion has been made ($\underline{34}$) that $C_3H_3^+$ does not react sequentially with acetylene to form small polycyclic ions, but rather directly with aromatic neutrals (benzene, toluene, methylnaphthalenes, indene) to form the initial polycyclic ions in one step. While these neutrals have only been seen as minor components in flames, their reactions with $C_3H_3^+$ may nonetheless be important since they would eliminate 3-5 sequential acetylene reactions, each with its possibility for branching into unreactive as well as reactive species. We have used ($\underline{35}$) electron impact on propargyl iodide to produce both cyclopropenylium and propargylium ions in our icr mass spectrometer and then studied their subsequent ion/molecule reactions with small aromatic neutral molecules. The results for both ionic structures are given in Table I, along with the expected rate coefficient calculated assuming an ion-induced dipole model ($\underline{36}$). One sees that as with

other hydrocarbon species, the propargylium ion reacts quite
rapidly with all of the neutrals, while the cyclopropenylium form
reacts slower or not at all. These results indicate that direct
reaction of $C_3H_3^+$ with cyclic neutrals may indeed be an important
channel in an ionic soot formation mechanism, since the rate
coefficients are quite high. However, importance of this channel
relative to sequential acetylene addition reactions cannot be
assessed until more of the rate coefficients, ionic structures,
and heats of formation have been determined for the reactions in
the latter scheme.

Table I. Rate Coefficients for Some Reactions of $C_3H_3^+$

$C_3H_3^+$ + Reactant Neutral ---→ Products

Reactant Neutral	Cyclic	Linear	Langevin
Acetylene (C_2H_2)	N.R.	12	11
Benzene (C_6H_6)	N.R.	15	16
Toluene (C_7H_8)	0.17	15-20	16
Naphthalene ($C_{10}H_8$)	0.17	7.0	13
1-Methylnaphthalene ($C_{11}H_{10}$)	N.R	5.7	--
2-Methylnaphthalene ($C_{11}H_{10}$)	0.21	1.6	--
Indene (C_9H_8)	4.4	18-20	--

All rate coefficients in cm^3/s x 10^{10}.
N.R. = no reaction or reaction less than cell loss.
-- = not calculated.

        The reactivity of the propargylium form and the non-
reactivity of the cyclopropenylium form of $C_3H_3^+$ toward acetylene
led to development of a new mass spectrometric technique for the
differentiation of structural isomers. In collaboration (37) with
Drs. Yost and Fetterolf in our department we have used reactive
collisions of low kinetic energy ions in the center quadrupole of
a triple quadrupole mass spectrometer to differentiate between the
two isomeric forms of $C_3H_3^+$. $C_3H_3^+$ ions were formed by electron
impact on propargyl chloride, bromide, and iodide, mass-selected
in the first quadrupole, then passed into the second quadrupole
with energies of 2-20 eV where they reacted with acetylene. Ions
resulting from these reactions, as well as unreacted $C_3H_3^+$ ions,
were then mass-analyzed in the third quadrupole. The intensity of
$C_5H_5^+$, formed by the reaction of the propargylium form of $C_3H_3^+$

with acetylene, increased slightly relative to unreacted $C_3H_3^+$ when propargyl bromide instead of chloride was used as precursor of $C_3H_3^+$; a marked increase was seen when propargyl iodide was used. These results are consistent with icr ion reactivity data that show electron impact upon propargyl iodide produces predominantly the linear, propargylium $C_3H_3^+$ ion, while use of the bromide or chloride produces almost exclusively the cyclopropenylium ion. Since quadrupole mass spectrometers have been employed in some flame sampling studies, our differentiation of ion structures with a triple quadrupole instrument suggests the possibility of actually determining the structures of selected ions sampled mass spectrometrically from flames. Such structure determination could serve to confirm, or further elucidate, an ionic soot nucleation pathway.

The number densities determined for $C_3H_3^+$ ions in several experimentally·studied flames are sufficiently high that use of laser induced fluorescence (lif) might be considered as a relatively non-intrusive probe for mapping the intensity distribution of these ions in the flame. However, some knowledge of the energies of the various excited states of the ions is necessary before a serious attempt at using lif can be made. The excited state energetics depend, of course, on the structure of the $C_3H_3^+$ isomer under consideration. No experimental determination of the excited states of either form of the ion considered here has been carried out. One rather complete theoretical calculation has been performed (38) on the cyclopropenylium ion, with the SCF LCAO-MO CI results indicating that the first excited singlet of the ion lies some 8.5 eV above the ground state. The excitation wavelength for lif studies on this structure would thus be ca. 146 nm, shorter than that currently available from commercial lasers. However, one might expect that the linear, propargylium form of the $C_3H_3^+$ ion would possess lower-lying excited states. No calculation of the excited state energetics of this structural isomer has been performed. In collaboration with Drs. Zerner and Edwards of the University of Florida Quantum Theory Project, we are currently investigating this problem. A spectroscopic INDO program (39) has been used to obtain excitation energies for the propargylium structure, and to confirm the earlier (38) results for the cyclopropenylium ion. Several low lying excited states, in the range 3-5 eV above the ground state, have been predicted, but all are symmetry forbidden to first order, and the forbiddeness is not broken by vibronic coupling. A second, more detailed set of calculations, in collaboration with Drs. Sabin and Oddershede, using a polarization propagator method (40) has produced results somewhat more promising than reported above. A transition requiring 5.36 ev energy (corresponding to a photon wavelength of 231 nm) with an oscillator strength of 0.14 has been predicted. This transition might be excited by laser radiation from a frequency doubled dye laser mixed with the fundamental output of a Nd:YAG pump laser. Thus lif detection of linear, propargylium

$C_3H_3^+$ ions in flames is a possibility, although interferences from other (possibly aromatic) species excited by light in this wavelength region would have to be eliminated.

$\underline{C_6H_5^+}$.   The $C_6H_5^+$ ion, although not a major ion in acetylene flames, has been seen in certain other studies (41), including those probing benzene flames. We also observed this ion in one of our early investigations (42) of sequential ion/molecule reactions in acetylene, where it was produced by the reaction of $C_4H_3^+$ ions with the parent neutral compound. Similar to the icr result discussed above for $C_3H_3^+$, the $C_6H_5^+$ ion was produced as a mixture of isomers, one reactive toward acetylene, and one unreactive. As with $C_3H_3^+$, the two most likely isomeric structures are a cyclic one **III**, the phenylium ion, and an acyclic ion **IV**.

III   IV

An ab initio (43) and a MINDO/3 (44) calculation have given some-what differing values for the heat of formation of the phenylium ion (280 vs. 245 kcal/mol, although see ref. 44 for a discussion which resolves this difficulty). Experimental ΔH_f values are in the range 266–270 kcal/mol (45, 46). Similarly, the open chain isomer **IV** has been calculated to have a heat of formation 17 and 16 kcal/mol higher than the phenylium by the ab initio and MINDO/3 calculations, respectively. The MINDO/3 results give substantially higher heats of formation for other isomeric forms.

We used both icr reactivity studies and collision-induced dissociation (cid) in a ZAB-2F double-focusing, reversed-geometry mass spectrometer to study the structures of $C_6H_5^+$ isomers (47). Both lines of investigation made use of the earlier report (48) that $C_6H_5^+$ ions possessing the phenylium structure could be formed by proton transfer to halobenzenes, followed by the loss of HX. Ions so formed in the icr mass spectrometer were found to be unreactive toward acetylene, indicating that in sequential ion/molecule reactions in acetylene, the phenylium $C_6H_5^+$ isomer is unreactive and the acyclic isomer **IV** is the reactive species. This conclusion was further borne out by an extensive series of cid experiments carried out in collaboration with Dr. J.E. Campana at the Naval Research Laboratory. $C_6H_5^+$ ions were formed in a variety of ways: direct electron impact on a number of neutral precursors; proton-assisted dehydrohalogenation reactions like those shown (48) to produce phenylium ions; and sequential acetylene ion/molecule reactions. The latter two formation schemes were carried out in the mass spectrometer's high-pressure chemical ionization source. Mass selected $C_6H_5^+$ ions were investigated using both charge stripping (49) reactions and fragmentation into $C_4H_2^+$ and $C_4H_3^+$ ions, where the theoretical calculations (44) predicted that the acyclic isomer should produce relatively more of the latter ion. Both classes of high energy processes verified that the reactive isomer produced by sequential ion/molecule reactions in acetylene has the acyclic structure **IV** and the unreactive isomer possesses the phenylium structure. Thus similar behavior (acyclic reactive, cyclic unreactive) has been found for both $C_3H_3^+$ and $C_6H_5^+$ ions.

$C_5H_5^+$. Reaction of the propargylium form of $C_3H_3^+$ with acetylene has been shown (33) to produce both $C_5H_5^+$ and $C_5H_3^+$. In flame sampling experiments the latter ion is usually more abundant (34) (although under certain conditions, the relative abundances of the two product ions are reversed (50)). Our work to date has concentrated on $C_5H_5^+$ because it can be formed in relatively high abundance in our icr mass spectrometer from a number of neutral precursors. There is substantial confusion, however, as to the relative stability of the several possible isomeric structures of this ion. Most theoretical studies (51) have concentrated on the cyclic structure, **V**,

V

while some experimental work was interpreted (52) in terms of both a cyclic and an acyclic isomer. However, a number of other structures, as seen in Figure 1, can be envisioned, and a few of even the more unlikely looking of these have been the subject of theoretical calculations (53). We have formed $C_5H_5^+$ ions from four neutral precursors and studied their reactions with various flame molecules. In addition, because of the lack of a consistent

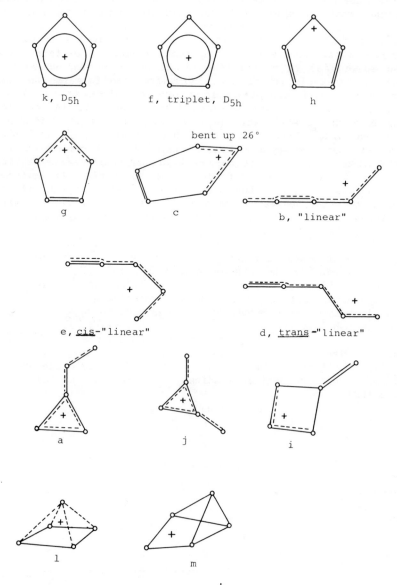

Figure 1. Some possible $C_5H_5^+$ isomeric structures.

theoretical treatment of the energetics of the many possible isomeric forms of $C_5H_5^+$, we have carried out MINDO/3 calculations on many of them, and more detailed ab initio calculations on three of them.

Two different primary ionization schemes were employed to form $C_5H_5^+$ ions. The first was conventional electron impact on cyclopentadiene, dicyclopentadiene, 1-penten-3-yne, and norbornadiene. The second involved charge transfer reactions in the icr analyzer cell from CO^+ to norbornadiene (54). Results (55) of these reactivity studies are given in Tables II - V. As can be seen, different neutral precursors produce isomers of differing reactivity, ranging from quite slow (ca. 3 x 10^{-12} cc/s in Table IV) to very high (ca. 5 x 10^{-10} in Table IV). Because of uncertainty in the identity of the various $C_5H_5^+$ structural isomers, we have labelled them simply A-D in the Tables. The reaction coefficients determined for isomer A are at the lower range of those which can be conveniently measured by the icr technique, and may represent non-reactive ion loss from the icr analyzer cell.

Table II.   Rate Coefficients for Some Reactions of $C_5H_5^+$ Ions
Formed from Cyclopentadiene

$C_5H_5^+$ + Reactant Neutral $\longrightarrow$ Products

	Ion Structure			
Reactant Neutral	A	B	C	D
Acetylene		0.20		
Ethylene (C_2H_4)		0.45		
Propane (C_3H_8)		0.32		
Methane (CH_4)		0.16		
1,3-Butadiene (C_4H_6)		0.62		
Allene (C_3H_4)		0.27		
Benzene		0.54		
Toluene		0.56		
Naphthalene		0.22		
1-Methylnaphthalene		0.18		
2-Methylnaphthalene		0.21		
Indene		0.39		

All rate coefficients in cm^3/s x 10^{10}.

Table III.   Rate Coefficients for Some Reactions of $C_5H_5^+$ Ions
             Formed from Dicyclopentadiene

$C_5H_5^+$ + Reactant Neutral ⟶ Products

Reactant Neutral	A	B	C	D
Acetylene			0.71	
Ethylene			0.49	
Propane			0.70	
1,3-Butadiene			0.72	
Allene			0.38	
Benzene			0.93	
Toluene			1.10	
Naphthalene			0.31	
1-Methylnaphthalene			0.33	
2-Methylnaphthalene			0.39	
Indene			0.46	

All rate coefficients in cm^3/s x 10^{10}.

Table IV.   Rate Coefficients for Some Reactions of $C_5H_5^+$ Formed
            from 1-penten-3-yne

$C_5H_5^+$ + Reactant Neutral ⟶ Products

Reactant Neutral	A	B	C	D
Acetylene	0.070			0.40
Ethylene	0.050			2.9
Propane	0.070			1.10
Methane	0.030			0.60
1,3-Butadiene	0.16			2.5
Allene	0.020			2.3
Benzene	0.40			3.4
Toluene	0.30			4.0
Naphthalene	0.030			1.0
1-Methylnaphthalene	N.R.			0.70
2-Methylnaphthalene	0.023			3.0
Indene	0.090			8.0

All rate coefficients in cm^3/s x 10^{10}.
N.R. = no reaction or reaction less than cell loss.

Table V.   Rate Coefficients for Some Reactions of $C_5H_5^+$ Ions
           Formed from Norbornadiene

$C_5H_5^+$ + Reactant Neutral ⟶ Products

Reactant Neutral	Ion Structure			
	A	B	C	D
Acetylene		0.20		0.53
Ethylene		--		--
Propane		0.23		2.7
Methane		--		--
1,3-Butadiene		0.63		2.2
Allene		--		--
Benzene		0.51		4.4
Toluene		--		--
Naphthalene		--		--
1-Methylnaphthalene		--		--
2-Methylnaphthalene		0.27		2.4
Indene		0.20		8.6

All rate coefficients in $cm^3/s \times 10^{10}$.

Isomers B and C, whose rate coefficients differ by less than a
factor of two in most cases, may actually be the same isomer
formed with differing degrees of internal excitation. However,
the substantial reactivity differences between A and B or C, and B
or C and D, lead us to the conclusion that at least three, and
possibly four, different isomeric structures are formed from these
precursors.

      To aid in isomeric identification in the $C_5H_5^+$ reactions, we
undertook a series of MINDO/3 calculations to determine the heats
of formation of the structures shown in Figure 1.   The MINDO/3
method has been demonstrated to give quite reasonable results for
geometries and heats of formation of both neutral and positively
charged hydrocarbon species (56).   We used a gradient geometry
optimization routine with a modified MINDO/3 program on our campus
Amdahl 450-V7 computer.   The results (55) are shown in Table VI,
where the lowest energy isomer is predicted to be the vinylcyclopropenylium ion, with several acyclic isomers next highest in
energy, and some of the cyclic species higher still.   MINDO/3 is
known to have several deficiencies, including underestimation of
small ring strain and prediction of the stability of triple bonds
ca. 5 kcal/mol too low.   Since the lowest heat of formation found
in our calculations was for a compound with a three-membered ring,
and since the acyclic isomers contain triple bonds, the MINDO/3

CHEMISTRY OF COMBUSTION PROCESSES

Table VI.   MINDO/3 Results for $C_5H_5^+$ Isomers

Structure*	ΔH_f (kcal/mol)
m	288
l, pyramid	270
k, D_{5h}, singlet	268
j, dimethylenecyclopropenylium	266
i, methylenecyclobutenylium	257
h	256
c, g	255
f, D_{5h}, triplet	253
e, cis-"linear"	252
d, trans-"linear"	247
b, "linear"	242
a, vinylcyclopropenylium	238

*See text and Figure 1 for structural details.

results are suspect. We have thus chosen three isomeric forms, a, b and c from Figure 1 and carried out ab initio calculations using a 4-31G basis set and a modified Gaussian 80 (57) package on our Quantum Theory Project VAX 11/780 computer. These more detailed calculations confirm the order of stability for the three isomers found using MINDO/3. We are currently using a fourth order many-body perturbation theory package (58) which includes single, double and quadruple excitations to see if the relative stabilities of the three isomeric forms chosen shift further. Preliminary results show that isomers a and b have essentially the same heat of formation, while the ΔH_f of c remains ca. 15 kcal/mol higher.

   The relative stability order predicted by our MINDO/3 calculations has led to a tentative assignment of $C_5H_5^+$ isomers, although this may have to be modified when final results of the more detailed quantum mechanical calculations discussed above are available. We assign the vinylcyclopropenylium structure (a) calculated to be lowest in energy, to the unreactive isomer A. The moderately reactive isomers B and C, which may be the same isomer with different amounts of internal energy, we assign to one or more of the acyclic forms similar to b in Figure 1. The highly reactive isomer, D, we assign the next higher energy cyclic structure, c, from Figure 1. Reactivity differences between the species A and B or C, and B or C and D are so great that we believe they are the result of the ions possessing different isomeric structures, and not just containing differing amounts of internal energy. If this structural assignment proves correct, it provides a different case from that which held for $C_3H_3^+$ and

$C_6H_5^+$. Now the cyclic isomer is less stable, and more reactive, with the acyclic less reactive, and a different, mixed cyclic and linear structure (the vinylcyclopropenylium ion) is the least reactive species.

A synthesis of both experimental and theoretical results generated by earlier workers and our group still seems insufficient to determine conclusively which isomeric structure should be assigned to the various reactive species which have been observed experimentally. The technique of photodissociation of gaseous ions trapped in an icr mass spectrometer has been used successfully for structural identification in a number of previous studies (59), and we hope to apply it to the present case. However, it is necessary to have some idea of the approximate spectra of the various $C_5H_5^+$ isomers which might be expected, in order to see if they are sufficiently different that structural differentiation might be accomplished. Also, some idea of the positions of various absorption bands is necessary in order to know which light sources (and most often which wavelengths of tunable lasers) must be used for the photodissociation. In collaboration with Drs. Zerner and Edwards of the UF Quantum Theory Project, we have theoretically predicted the spectra of four $C_5H_5^+$ isomers, using the INDO spectral prediction program (39). Results are shown in Figure 2 for four of the possible isomeric structures given in Figure 1. One sees that at least one is predicted to have a moderately intense absorption band in the near ultraviolet region, which should be accessible by either excimer-pumped or Nd:YAG-pumped dye lasers (but unfortunately not by our existing flashlamp-pumped dye laser). With an improved laser capability (or perhaps using multiple photon ir laser-induced dissociation (60) with our cw CO_2 laser), we should be able to assign more completely the $C_5H_5^+$ isomeric structures.

Conclusion

Studies of the structure and reactivities of the three ions discussed in this article have continued to bear out the possibility of an ionic mechanism for soot formation. The reaction rate coefficients are certainly rapid and in general larger than those used by Olson and Calcote (10) in their model, at least for certain isomeric forms of the ions and at the temperatures of ca. 325 K used in our work. However, until more sophisticated flame-sampling mass spectrometers are employed, the exact isomeric form of the many ions seen in flames cannot be known, and thus cannot be related directly to our work. As is to be expected, details of the reactivity or nonreactivity of various isomers vary as one moves from one ionic species to another. Thus in some cases ($C_5H_5^+$) acyclic forms are less reactive, while the cyclic isomers react rapidly, while in others ($C_3H_3^+$, $C_6H_5^+$) the opposite is the case. Those channels where the cyclic ions are less reactive suggest an opportunity for the formation of cyclic neutrals in

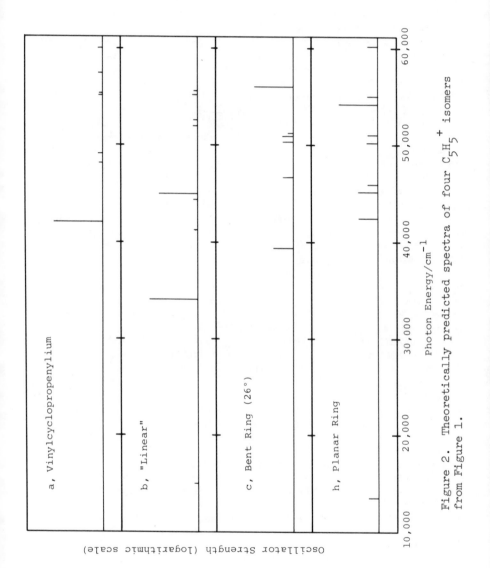

Figure 2. Theoretically predicted spectra of four $C_5H_5^+$ isomers from Figure 1.

flames by ion recombination with electrons or hydride transfer reactions. Our findings on the reactivity of $C_3H_3^+$ ions with aromatic neutrals suggest that this may be an important channel in the formation of larger cyclic ions, as postulated earlier (34).

While isolated laboratory experiments involving reactions which may be important in flames provide information under controlled conditions, these conditions may be somewhat removed from those which are found in flames. Clearly our reaction rate determinations should be extended to higher temperatures, which may cause the association reactions seen to have lower rates, due to dissociation of the ion/molecule reaction complexes formed. Also, studies at higher pressures should be performed, where there is increased opportunity for collisional stabilization of the collision complexes. As data from work in our laboratories and those of others accumulate, they can be used to refine computational models such as those already reported (10), in order to more fully test the proposed ionic soot formation mechanism.

Acknowledgments

Acknowledgment is made to the Donors of The Petroleum Research Fund, administered by the American Chemical Society, for the partial support of this research.

Literature Cited

1. Lawton, J.; Weinberg, F.J. "Electrical Aspects of Combustion"; Clarendon Press: Oxford, 1969.
2. Knewstubb, P.F.; Sugden, T.M. Nature 1958, 181, 1261.
3. Calcote, H.F. in "Eighth Symp. (Int'l.) on Combustion"; Williams and Wilkins Co.: Baltimore, 1962; p. 184.
4. Hayhurst, A.N.; Kittelson, D.B. Combustion and Flame 1978, 31, 37.
5. Delfau, J.L.; Michaud, P.; Barassin, A. Combust. Sci. and Tech. 1979, 20, 165.
6. Olson, D.B.; Calcote, H.F. in "Eighteenth Symp. (Int'l.) on Combustion"; The Combustion Institute: Pittsburgh, 1981; p. 453.
7. Goodings, J.M.; Bohme, D.K.; Sugden, T.M. in "Sixteenth Symp. (Int'l.) on Combustion"; The Combustion Institute, Pittsburgh, 1977; p. 891.
8. Goodings, J.M.; Bohme, D.K.; Ng, C.-W. Combustion and Flame 1979, 36, 27.
9. Calcote, H.F. Combustion and Flame 1981, 42, 215.
10. Olson, D.B.; Calcote, H.F. in "Particulate Carbon Formation during Combustion"; Siegla, D.C.; Smith, G.W., Eds.; Plenum: New York, 1981; p. 177.
11. Goodings, J.M.; Tanner, S.D.; Bohme, D.K. Can. J. Chem. 1982, 60, 2766.

12. Baldeschwieler, J.D.; Woodgate, S.S. Accts. Chem. Res. 1971, 4, 114.
13. Beauchamp, J.L. Ann. Rev. Phys. Chem. 1971, 22, 527.
14. Henis, J.M.S. in "Ion-Molecule Reactions, Vol. 2"; Franklin, J.L., Ed.; Plenum: New York, 1972; p. 395.
15. Lehman, T.A.; Bursey, M.M. "Ion Cyclotron Resonance Spectrometry"; Wiley: New York, 1976.
16. McIver, R.T., Jr., Rev. Sci. Instrum. 1978, 49, 111.
17. Ausloos, P.; Eyler, J.R.; Lias, S.G. Chem. Phys. Lett. 1975, 30, 21.
18. Aue, D.H.; Bowers, M.T. in "Gas Phase Ion Chemistry, Vol. 2"; Bowers, M.T., Ed.; Academic: New York, 1979, p. 2.
19. Bartmess, J.E.; McIver, R.T., Jr., in "Gas Phase Ion Chemistry, Vol. 2"; Bowers, M.T., Ed.; Academic: New York, 1979; p. 88.
20. Gross, M.L.; Russell, D.H.; Aerni, R.J.; Bronczyk, S.A. J. Am. Chem. Soc. 1977, 99, 3603.
21. Jackson, J.A.A.; Lias, S.G.; Ausloos, P. J. Am. Chem. Soc. 1977, 99, 7515.
22. Ausloos, P. J. Am. Chem. Soc. 1981, 103, 3931.
23. Yost, R.A.; Enke, C.G. Anal. Chem. 1979, 51, 1251A.
24. Slayback, J.R.B.; Story, M.S. Ind. Res. Dev., February 1981, 128.
25. Fetterolf, D.D.; Yost, R.A. Mass Spectrom. Rev. 1983, 2, 1.
26. Brotherton, H.D.; Yost, R.A. Anal. Chem. 1983, 55, 549.
27. Morgan, R.P.; Beynon, J.H.; Bateman, R.H.; Green, B.N. Int. J. Mass. Spectrom. Ion Phys. 1978, 28, 171.
28. Cooks, R.G. in "Collision Spectroscopy"; Cooks, R.G., Ed.; Plenum: New York, 1978.
29. Radom, L.; Hariharan, P.C.; Pople, J.A.; Schleyer, P.v.R. J. Am. Chem. Soc. 1976, 98, 10.
30. Lossing, F.P. Can. J. Chem. 1972, 50, 3973.
31. Knewstubb, P.F.; Sugden, T.M. in "Seventh Symp. (Int'l) on Combustion"; Butterworths: London, 1959, p. 247.
32. Ausloos, P.J.; Lias, S.G. J. Am. Chem. Soc. 1981, 103, 6505.
33. Smyth, K.C.; Lias, S.G.; Ausloos, P. Combust. Sci. and Tech. 1982, 28, 147.
34. Michaud, P.; Delfau, J.L.; Barrasin, A. in "Eighteenth Symp. (Int'l) on Combustion"; The Combustion Institute: Pittsburgh, 1981, p. 433.
35. Brill, F.W.; Baykut, M.G.; Eyler, J.R., manuscript in preparation.
36. Giomousis, G.; Stevenson, D.P. J. Chem. Phys. 1958, 29, 294.
37. Fetterolf, D.D.; Yost, R.A.; Eyler, J.R. Org. Mass Spectrom., in press.
38. Takada, T.; Ohno, K. Bull Chem. Soc. Japan 1979, 52, 334.
39. Ridley, J.; Zerner, M. Theor. Chim. Acta 1973 32, 111.
40. Oddershede, J. in "Advances in Quantum Chemistry, Vol. II"; Lowdin, P.O., Ed.; Academic: New York, 1978; p. 275.

41. Vinckier, C.; Gardner, M.P.; Bayes, K.D. in "Sixteenth Symp. (Int'l.) on Combustion"; The Combustion Institute: Pittsburgh, 1976, p. 881.
42. Brill, F.W.; Eyler, J.R. J. Phys. Chem. 1981, 85, 1091.
43. Dill, J.D.; Schleyer, P.v.R.; Brinkley, J.S.; Seeger, R.; Pople, J.A.; Haselbach, E. J. Am. Chem. Soc. 1976, 98, 5428.
44. Tasaka, M.; Ogata, M.; Ichikawa, H. J. Am. Chem. Soc. 1981, 103, 1885.
45. McMahon, T.B. Ph.D. Thesis, California Institute of Technology, Pasadena, 1973.
46. Rosenstock, H.M.; Larkins, J.T.; Walker, J.A. Int. J. Mass. Spectrom. Ion Phys. 1973, 11, 309.
47. Eyler, J.R.; Campana, J.E. Int. J. Mass. Spectrom. Ion Phys., in press.
48. Speranza, M.; Sefcik, M.D.; Henis, J.M.S.; Gaspar, P.P. J. Am. Chem. Soc. 1977, 99, 5583.
49. Ast, T. in "Advances in Mass Spectrometry, 8B"; Quayle, A., Ed.; Heyden & Son, Ltd.: London, 1980; p. 555.
50. Olson, D.B., personal communication.
51. Borden, W.T.; Davidson, E.R. J. Am. Chem. Soc. 1979, 101, 3771.
52. Occolowitz, J.L.; White, G.L. Aust. J. Chem. 1968, 21, 997.
53. Stohrer, W.D.; Hoffman, R. J. Am. Chem. Soc. 1972, 94, 1661.
54. Buckley, T.J. Ph.D. Thesis, University of Florida, Gainesville, 1982 and Lias, S.G.; Ausloos, P.; Buckley, T.J., unpublished results.
55. Brill, F.W. Ph.D. Thesis, University of Florida, Gainesville, 1983.
56. Bingham, R.C.; Dewar, M.J.S.; Lo, D.H. J. Am. Chem. Soc. 1975, 97, 1285.
57. Available from Quantum Chemistry Program Exchange, Department of Chemistry, Indiana University.
58. Bartlett, R.J.; Purvis, G.D. Phys. Scr. 1980, 21, 255.
59. Dunbar, R.C. in "Gas Phase Ion Chemistry, Vol. 2"; Bowers, M.T., Ed.; Academic: New York, 1979; Ch. 14.
60. Bomse, D.S.; Berman, D.W.; Beauchamp, J.L. J. Am. Chem. Soc. 1981, 103, 3967.

RECEIVED November 30, 1983

FUEL NITROGEN

Nitrogen Chemistry in Flames
Observations and Detailed Kinetic Modeling

ANTHONY M. DEAN, MAU-SONG CHOU, and DAVID STERN

Corporate Research—Science Laboratories, Exxon Research and Engineering Company, Linden, NJ 07036

Spatially resolved concentration measurements of
NH, NH_2, NH_3, NO and OH in atmospheric pressure
ammonia flames are compared to predictions obtained
with a one-dimensional flame algorithm using a
detailed reaction mechanism. Several reactions
were observed to be equilibrated, and this information was used to obtain estimates of the oscillator
strength for NH_2 as well as the heat of formation
of NH. Use of a conventional mechanism of ammonia
oxidation predicted concentration profiles in marked
disagreement with the observations. However, it was
possible to obtain much more satisfactory fits
to the data by including reactions between various
NH_i (i = 1-2) species to form N-N bonds; these
adducts could then decompose to form ultimately N_2.
These good fits were obtained with rate constants
estimated from unimolecular decomposition theory
and used with no adjustments.

Recent advances in both laser diagnostic instrumentation and
computer modeling algorithms have provided kineticists with
exciting new opportunities for characterization of complex chemical systems. In this paper we describe our use of these tools
to elucidate the kinetics of nitrogen species in flames. Our
efforts have focused upon ammonia oxidation since it can be
shown to play a key role in terms of NO production in fuel-bound-
nitrogen flames (1) as well as NO reduction via NH_3 addition to
flue gas in the Thermal $DeNO_x$ process (2,3). Although there
have been a host of previous studies of ammonia oxidation in
both shock tubes (4) and flames (5), considerable ambiguity
remains with respect to the details of the mechanism. We have
attempted to remove some of these uncertainties by measuring
absolute concentrations of both stable species and reactive
intermediates through the flame front region and comparing these
concentration-distance profiles to those obtained by numerically

0097-6156/84/0249-0071$06.00/0

solving the 1-D flame equations. A discussion of the measurement
techniques is followed by the mechanistic analysis. Portions of
this material have been covered in more detail in our recent
papers (6-8).

Experimental Arrangement and Observations

Figure 1 shows a schematic of the apparatus. This has been
described in detail in Refs. 6 and 7; only an overview is presented here. The 3.8 cm diameter burner is designed to produce a
flat horizontal flame front so that one can map out the flame
chemistry by measuring species concentrations at various vertical
distances above the burner surface. This geometry allows one the
considerable simplification of utilizing a one-dimensional flame
code for the kinetic analysis. The laser beam is focused to give
a constant beam diameter of $\sim$0.15 mm across the flame. Again
this simplifies the interpretation of the line-of-sight absorption
measurements. Furthermore, this small diameter permits good
spatial resolution of the flame front region. In our atmospheric
pressure flames, the flame front was less than 1 mm thick.
Averaging techniques were used for collection of both absorption
and fluorescence data, with 100 laser pulses per data point being
typical. The large temperature gradients near the burner surface
caused some beam steering, but direct measurement showed this
deflection to be negligible, i.e., less than 0.08 mm, at distances
larger than 0.3 mm above the surface.

The gas mixtures fed to the burner were prepared in a stainless steel manifold using electronic flow controllers. A range
of rich ammonia flames was studied in which the fuel equivalence
ratio ranged from 1.28 to 1.81. Flame temperatures were measured
with Pt/Pt-13%Rh thermocouples. The bead diameter was only 0.12
mm so that the radiation correction was only 80 K.

OH, NH, NH_2, and NH_3 were measured in absorption while NO
was measured in fluorescence. The absorption measurements were
reduced to concentrations via curve-of-growth techniques while
the NO measurements were calibrated against absorption in a lean
ammonia flame where the NO concentration was higher. It was
assumed that the extent of fluorescence quenching was the same in
the rich and lean flames (7). Sufficient spectroscopic information was available for all but NH_2 to allow absolute concentration
assignments. NH_2 data could only be obtained as $[NH_2] \cdot f_i$, where
f_i was the unknown oscillator strength.

Since the OH absorption measurements were made on individual
rotational lines, it was possible to obtain $[OH]_J$. Assuming a
Boltzmann distribution for the rotational energy levels, one can
obtain the rotational temperature from a plot of $\ln[OH]_J$ versus
E_{ROT}. An example of such data is shown in Figure 2.

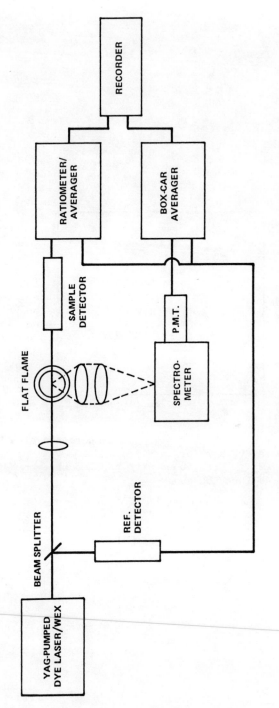

Figure 1. Schematic diagram of the experimental arrangements for laser absorption and laser-induced fluorescence. (Reproduced with permission from Ref. 7. Copyright 1983, J. Chem. Phys.)

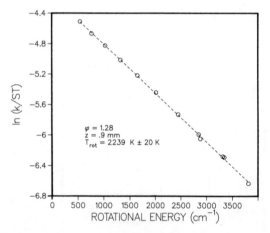

Figure 2. Measurement of OH rotational temperature at a height of 0.9 mm above the burner for an ammonia flame with an equivalence ratio of 1.28. (Reproduced with permission from Ref. 6. Copyright 1982, J. Chem. Phys.)

Results and Discussion

NH$_2$ Oscillator Strength.   As mentioned earlier, lack of a reliable oscillator strength for NH$_2$ prevented assignment of absolute
NH$_2$ concentrations.   However, we were able to combine absolute
measurements of OH, NH, and NH$_3$ with relative measurements of NH$_2$
to demonstrate that the reactions

$$NH_3 + OH = NH_2 + H_2O \qquad (1)$$

$$NH_2 + OH = NH + H_2O \qquad (2)$$

were partially equilibrated.   This fact allowed us to calculate
absolute NH$_2$ concentrations and thus obtain f_i. The use of two
reactions not only allowed a consistency check, but also served
to provide an estimate for $\Delta H_f°(NH)$.
      The approach used assumes that H$_2$O rapidly assumes an equili-
brium concentration in flames. With H$_2$O fixed in this way, one
can explicitly verify whether or not Reactions 1 and 2 are equilibrated by computing the ratios:

$$[NH_2] \cdot f_i / ([NH_3] \cdot [OH]) \qquad (3)$$

$$[NH] / ([NH_2] \cdot f_i \cdot [OH]) \qquad (4)$$

Representative values of (3) and (4) are tabulated in Table I.
Although the individual concentrations vary widely, each of these
ratios is constant within experimental error at different heights
above the burner, indicating that each of the reactions is equilibrated.   Calculation of K_{eq} is more uncertain for Reaction 2
because of the uncertain heat of formation of NH.   Given this
uncertainty, one can vary $\Delta H_f°(NH)$ until f_i obtained from
Reaction 2 agrees with that obtained from Reaction 1.   Such an
approach yields $f_i = 5.05 \times 10^{-5}$ with $\Delta H_f°(NH) = 89$ kcal/mole.
This value of $\Delta H_f°(NH)$ is consistent with the JANAF value of
90 ± 4 kcal/mole [9] but higher than the 84.2 ± 2.3 kcal/mole proposed
by Piper [10].   Since there are additional uncertainties in
$\Delta H_f°(NH_2)$ and the absolute concentration of NH$_3$, NH, and OH, these
values of f_i and $\Delta H_f°(NH)$ cannot be taken as definitive assignments, but they do represent significant improvements over earlier
work.

Rotational Excitation of OH.   One of the most surprising aspects
of our data was the observation of rotationally hot OH in the
flame front of $\phi = 1.28$ and $\phi = 1.50$ flames.   Rotational
temperatures ~ 200 K higher than radiation corrected thermocouple
measurements were observed; these were not expected since rotational energy transfer is so fast at atmospheric pressure.   Such
excitation was not observed beyond the flame front in any of our
ammonia flames and not even within the flame front of a methane

Table I. Partial Equilibrium in Ammonia Flames

$\phi = 1.28$ ($NH_3/O_2/N_2 = 4.79/2.81/1.00$)

z/mm[a]	T/K	$[OH]/10^{14}cm^{-3}$ [b]	$[NH]/10^{14}cm^{-3}$ [b]	$[NH_2]\cdot f_i/10^{11}cm^{-3}$ [b]	$[NH_3]/10^{17}cm^{-3}$ [b]	$\dfrac{[NH_2]\cdot f_i}{[OH][NH_3]}/10^{-22}cm^{+3}$ [c]	$\dfrac{[NH]}{[NH_2]\cdot f_i[OH]}/10^{-12}cm^{+3}$
1.0	1938	14.3	5.28	2.54	2.04	8.71	1.45
1.5	1941	12.0	2.56	1.40	1.40	8.33	1.52
2.0	1938	10.0	1.37	0.91	1.11	8.20	1.50
2.5	1935	9.22	0.986	0.69	0.90	8.32	1.55
3.0	1929	8.46	0.724	0.54	0.79	8.08	1.58

[a] Distance above burner surface.

[b] Reference 6.

[c] Reference 8.

flame. It appeared that the source of this excited OH was an
exothermic reaction which was unique to the flame front region of
the ammonia flame. A rate analysis of the mechanism which is
discussed later indicates two reactions which satisfy these
criteria:

$$NH_2 + O \rightarrow NH + OH \qquad \Delta H = -14 \text{ kcal/mole}$$

$$NH + NO \rightarrow N_2 + OH \qquad \Delta H = -96 \text{ kcal/mole}$$

Both of these reaction rates peak in the flame front region, and
both have the potential to produce vibrationally excited OH.
Since vibrational relaxation is much slower than rotational
relaxation, appreciable quantities of vibrationally excited OH
could be formed. The observed rotational excitation could result
from cascading of this excitation into high rotational levels of
the ground vibrational level during the relaxation process.

Detailed Kinetic Modeling. Recent advances in computation techniques (11) have made it much easier to compute concentration-
distance profiles for flame species. The one-dimensional isobaric
flame equations are solved via a steady state solution using
finite difference expressions. An added simplification is that
the energy equation can be replaced with the measured temperature
profile. In the adaptive mesh algorithm, the equations are first
solved on a relatively coarse grid. Then additional grid points
could be included if necessary, and the previous solution inter-
polated onto the new mesh where it served as the initial solution
estimate. This process was continued until several termination
criteria were satisfied.

The starting point in development of an ammonia flame mechanism was a mechanism previously used to model ammonia oxidation
in a flow tube near 1300 K (3). Additional reactions were added
that were thought to be important at the higher flame temperatures.
Calculations with this mechanism produced profiles in marked disagreement with our data. The predictions were slower than
observed; decay of NH_i species was much too slow, and OH peaked
too late by about 2.5 mm. To make matters worse, far too much NO
was formed. The NO problem was especially troublesome in that
attempts to increase the rate of NH_i decay only served to produce
even more NO, since NO was the primary decay channel for the NH_i
species. A possible resolution of this dilemma involves reactions
of the NH_i species with each other to form N-N bonds. These
complexes could then split off H atoms to ultimately form N_2.
In this way one could achieve an overall faster decay of NH_i
without producing more NO. Indeed, calculations using such a
mechanism showed much better agreement with our data. Table II
lists the "conventional" mechanism as well as the additional
$NH_i + NH_i$ reactions which dramatically improve the fit. Figure 3

Table II.   Mechanism for Rich Ammonia Flames

		$k = AT^n \exp(-E/RT)$			
	Reaction	A	n	E(kcal/mole)	Comments
	UPDATED FLOW REACTOR MECHANISM				
1.	$NH_3+M=NH_2+H+M$	4.80E+16	0.	93929.	Ref. 3
2.	$NH_3+H=NH_2+H_2$	2.46E+13	0.	17071.	Ref. 3
3.	$NH_3+O=NH_2+OH$	1.50E+12	0.	6040.	Ref. 3
4.	$NH_3+OH=NH_2+H_2O$	3.26E+12	0.	2120.	Ref. 3
5.	$NH_2+O=NH+OH$	2.00E+13	0.	1000.	Ref. 3
6.	$NH_2+OH=NH+H_2O$	3.00E+10	0.68	1290.	Ref. 3
7.	$NH_2+H=NH+H_2$	1.00E+12	0.5	2000.	Ref. 8
8.	$NH_2+O_2=HNO+OH$	5.10E+13	0.	30000.	Ref. 3
9.	$NH_2+NO=NNH+OH$	6.10E+19	−2.46	1866.	Ref. 3
10.	$NH_2+NO=N_2+H_2O$	9.10E+19	−2.46	1866.	Ref. 3
11.	$NH_2+NO=N_2O+H_2$	5.00E+13	0.	24800.	Ref. 15
12.	$NH_2+HNO=NH_3+NO$	1.75E+14	0.	1000.	Ref. 3
13.	$NH_2+NNH=N_2+NH_3$	1.00E+13	0.	0.	Ref. 3
14.	$NH+O_2=HNO+O$	6.00E+12	0.	3400.	Ref. 8
15.	$NH+NO=N_2+OH$	1.20E+13	0.	0.	Ref. 8
16.	$NH+OH=N+H_2O$	5.00E+11	0.5	2000.	Ref. 16
17.	$NH+OH=HNO+H$	5.00E+11	0.5	2000.	Ref. 16
18.	$NH+H=N+H_2$	3.00E+13	0.	0.	Ref. 1
19.	$NH+O=NO+H$	6.30E+11	0.5	0.	Ref. 16
20	$NH+O=N+OH$	6.30E+11	0.5	8000.	Ref. 16
21.	$NH+N=N_2+H$	6.30E+11	0.5	0.	Ref. 16
22.	$HNO+M=NO+H+M$	1.86E+16	0.	48680.	Ref. 3
23.	$HNO+OH=NO+H_2O$	3.60E+13	0.	0.	Ref. 3
24.	$HNO+H=NO+H_2$	4.80E+12	0.	0.	Ref. 16
25.	$HNO+O=NO+OH$	5.00E+11	0.5	0.	Ref. 16
26.	$HNO+N=NO+NH$	1.00E+11	0.5	2000.	Ref. 16
27.	$HNO+N=H+N_2O$	5.00E+10	0.5	3000.	Ref. 16
28.	$NNH+M=N_2+H+M$	1.50E+15	0.	35000.	Ref. 8
29.	$NNH+OH=N_2+H_2O$	3.00E+13	0.	0.	Ref. 3
30.	$NNH+NO=N_2+HNO$	2.50E+12	0.	0.	Ref. 3
31.	$N+NO=N_2+O$	1.60E+13	0.	0.	Ref. 16
32.	$N+O_2=NO+O$	6.40E+09	1.0	6300.	Ref. 16
33.	$N+OH=NO+H$	6.30E+11	0.5	0.	Ref. 16
34.	$N_2O+M=N_2+O+M$	2.70E+14	0.	54100.	Ref. 17
35.	$N_2O+H=NH+NO$	3.80E+14	0.	34500.	Ref. 18
36.	$N_2O+H=N_2+OH$	7.60E+13	0.	15100.	Ref. 19
37.	$N_2O+O=N_2+O_2$	1.00E+14	0.	28020.	Ref. 19
38.	$N_2O+O=NO+NO$	1.00E+14	0.	28020.	Ref. 19
39.	$H_2+OH=H_2O+H$	2.20E+13	0.	5150.	Ref. 3
40.	$H+O_2=OH+O$	3.70E+17	−1.0	17500.	Ref. 20
41.	$O+H_2=OH+H$	1.80E+10	1.0	8900.	Ref. 3
42.	$H+HO_2=OH+OH$	2.50E+14	0.	1900.	Ref. 3
43.	$O+HO_2=O_2+OH$	4.80E+13	0.	1000.	Ref. 3
44.	$OH+HO_2=H_2O+O_2$	5.00E+13	0.	1000.	Ref. 3

Continued

Table II. Mechanism for Rich Ammonia Flames (Continued)

Reaction	A	n	E(kcal/mole)	Comments
45. $OH+OH=O+H_2O$	6.30E+12	0.	1090.	Ref. 3
46. $HO_2+NO=NO_2+OH$	3.43E+12	0.	-260.	Ref. 3
47. $H+NO_2=NO+OH$	3.50E+14	0.	1500.	Ref. 3
48. $O+NO_2=NO+O_2$	1.00E+13	0.	600.	Ref. 3
49. $H+O_2+M=HO_2+M$ $H_2O/21.7***$	1.50E+15	0.	-995.	Ref. 3
50. $NO_2+M=NO+O+M$	1.10E+16	0.	66000.	Ref. 3
51. $O+O+M=O_2+M$	1.38E+18	-1.0	340.	Ref. 3
52. $H+H+M=H_2+M$ $H_2/2.5/$ $H_2O/15./$	3.60E+16	-0.6	0.	Ref. 21
53. $H+OH+M=H_2O+M$ $H_2/2.5/$ $H_2O/15./$	8.80E+21	-2.0	0.	Ref. 21

INCLUSION OF NH + NH REACTIONS

Reaction	A	n	E(kcal/mole)	Comments
54. $NH_2+NH_2=N_2H_3+H$	1.00E+13	0.	16100.	Ref. 8
55. $NH_2+NH_2=N_2H_4$	5.00E+12	0.	0.	Ref. 8
56. $N_2H_4+H=N_2H_3+H_2$	1.00E+12	0.5	2000.	Same as $H+NH_2$
57. $N_2H_4+OH=N_2H_3+H_2O$	3.00E+10	0.68	1290.	Same as $OH+NH_2$
58. $N_2H_4+O=N_2H_3+OH$	2.00E+13	0.	1000	Same as $O+NH_2$
59. $N_2H_3=N_2H_2+H$	1.20E+13	0.	58000.	Ref. 8
60. $N_2H_3+H=N_2H_2+H_2$	1.00E+12	0.5	2000.	Same as $H+NH_2$
61. $N_2H_3+OH=N_2H_2+H_2O$	3.00E+10	0.68	1290.	Same as $OH+NH_2$
62. $N_2H_3+O=N_2H_2+OH$	2.00E+13	0.	1000.	Same as $O+NH_2$
63. $N_2H_2=NNH+H$	3.40E+12	0.	65000.	Ref. 8
64. $N_2H_2+H=NNH+H_2$	1.00E+12	0.5	2000.	Same as $H+NH_2$
65. $N_2H_2+OH=NNH+H_2O$	3.00E+10	0.68	1290.	Same as $OH+NH_2$
66. $N_2H_2+O=NNH+OH$	2.00E+13	0.	1000.	Same as $O+NH_2$
67. $NH+NH=NNH+H$	5.00E+13	0.	0.	Ref. 8
68 $NH+NH_2=N_2H_2+H$	5.00E+13	0.	0.	Ref. 8

NOTE: Rate constant units are mole, cm, sec,K.

*** i.e., rate constant increased by a factor of 21 for H_2O as the third body.

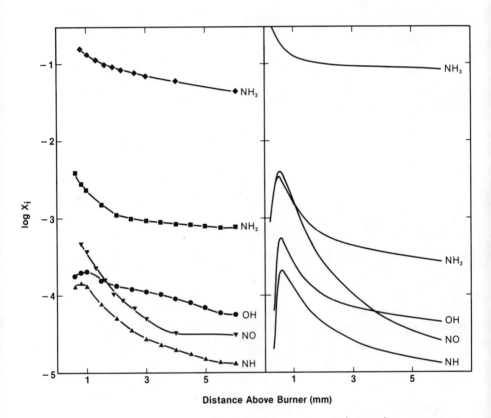

Figure 3. Comparison of (left) observed and (right) calculated profiles for an ammonia flame with an equivalence ratio of 1.50.

compares these predictions with the expanded mechanism to our observations for the ϕ = 1.50 flame.

Both calculations and observations contain potential sources of error, and detailed comparisons should be made with these in mind. The concentration measurements of NH and OH are probably accurate to ±20%, except that OH in the flame front region could be low if there were appreciable vibrational excitation. NO measurements were calibrated via absorption (7) and should also be accurate to ±20%. NH3 is more uncertain, ±30%, since there is a larger uncertainty in the extinction coefficient used (12). NH2 is probably only accurate to within a factor of two because of the uncertainties in f_i. Uncertainty is introduced into the calculations by the uncertain heats of formation of NH and NH2. With these uncertainties, major emphasis should be placed upon comparison of the shapes of the concentration profiles.

Using these guidelines, the comparisons in Figure 3 are generally quite satisfactory. Note that NH is properly described in terms of both profile shape as well as absolute concentration. The calculated NO profile near the burner is too high, but the overall decay seems to be reasonably well described. This discrepancy close to the burner surface can be resolved by using a larger rate constant for the reaction

$$NH + NO \rightarrow N_2 + OH$$

However, considering the uncertainty in absolute concentrations, it was felt that this was insufficient justification for use of a higher value. OH is also properly predicted, with the exception of the region near the peak, and this discrepancy may well be a manifestation of vibrationally excited OH as described earlier. NH2 can be seen to have the proper shape, and the absolute concentration predictions are probably acceptable considering the large error bars here. The predicted ammonia decay rate at larger distances above the burner is somewhat slower than observed. However, even in this case where the fit leaves something to be desired, it is vastly improved over what it was for the case where the NH_i + NH_j reactions were excluded.

In general, the fits illustrated in Figure 3 are similar to those observed at the other two equivalence ratios (8). A particularly encouraging aspect of the calculations at all three equivalence ratios is that they properly predict the variation of both peak height and peak location of the radical species with respect to changes in the equivalence ratio. Thus, it appears that the mechanism given in Table II properly accounts for most of our extensive data base on rich ammonia flames. Although it is impossible to prove that this is the correct mechanism, the proper prediction of so many species over a range of conditions strongly suggests that the scheme used is a reasonable approximation to reality. It is evident that the NH_i + NH_i reactions are the key ingredient to obtaining this good fit. As

mentioned earlier, omission of these reactions led to predictions
which bore no resemblance to our data (8).

Figure 4 outlines the important reactions of the nitrogen
species in rich ammonia flames. The most important reactions
producing N_2 are NNH dissociation and NH + NO. In turn, most of
the NNH is produced from N_2H_2 dissociation which is produced via
the reaction NH + NH_2. Other NH_i + NH_i reactions are less important. Thus, the NH_i + NH_i reactions are primarily responsible
for N_2 production; it is now clear why omission of these channels
led to such marked changes in the predicted profiles. One would
not expect such changes in predicted profiles in lean ammonia
flames; here $[NH_i]$ would be sufficiently low that the NH_i + NH_i
reactions could safely by omitted.

Figure 4 also outlines a possible reason for the continuing
controversy as to the identity of the NH_i species which was
responsible for the NH_i + NO reaction when examining NO production
in the combustion of nitrogen doped fuels. An analysis of the
reactions in Figure 4 indicates that the relative concentration
of NH_i species will vary with conditions. In this work where
large quantities of NH and NH_2 are present, N_2 production occurs
via NH_i + NH_i as discussed above. However, in the more typical
case of small amounts of nitrogenous diluent, the NH_i concentrations should be too low for these reactions to be significant.
There the relative population of the NH_i species will be governed
by the competition between the various NH_i + NO reactions and the
hydrogen abstractions:

$$X + NH_i \rightarrow HX + NH_{i-1}$$

Larger radical concentrations (X) would favor higher concentrations of N atoms; then N + NO would be important. Lower radical
concentrations would tend to increase the importance of NH + NO.
Finally, much lower temperatures, i.e., 1200 K for the Thermal
DeNO$_x$ process, would make NH_2 + NO a key reaction.

Estimation of Rate Constants for NH_i + NH_i and N_2H_i Dissociation.
The keys to the success of the mechanism are the NH_i + NH_i
reactions as well as the subsequent unimolecular dissociation
of the N_2H_i species formed from these reactions. A typical
scheme is the following:

$$NH + NH_2 \underset{k_r}{\overset{k_f}{\rightleftharpoons}} H_2NNH^{\ddagger} \overset{k_H}{\longrightarrow} HNNH + H$$

$$\Big\downarrow k_s[M] \qquad\qquad \Big\downarrow k_d$$

$$H_2NNH \qquad\qquad NNH + H$$

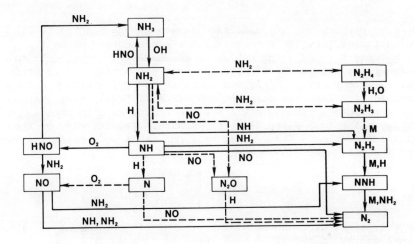

Figure 4. Important reactions of nitrogen species in rich ammonia flames.

This is written in the mechanism as the following sequence:

$$NH + NH_2 = N_2H_2 + H \tag{68}$$

$$N_2H_2 = NNH + H \tag{63}$$

with

$$k_{68} = k_H \left\{ \frac{k_f}{k_H + k_r + k_s M} \right\}$$

and

$$k_{63} = k_D = k_D^\infty \cdot I$$

One thus obtains the apparent rate constants k_{68} and k_{63} by evaluation of the rate constants for the elementary steps (k_f, k_r, k_H, and k_s) and using tables of the Kassel integral (13) to estimate the degree of fall-off (I) from the limiting high pressure rate constant, k_D^∞. k_f can be taken to be the high pressure recombination rate constant; k_s is the collisional stabilization rate constant; k_r and k_H are the unimolecular rate constants corresponding to N-N and N-H bond fission, respectively, of the collision complex. These decay rate constants were estimated from the RRK expression.

$$k = A \left[\frac{\varepsilon - \varepsilon^*}{\varepsilon} \right]^{S-1}$$

where A is the preexponential factor, $\varepsilon - \varepsilon^*$ is the amount of vibrational energy in excess of that required to break the bond of interest, ε is the total energy of the complex, and S is the effective number of oscillators (14). S is computed via the relation

$$S \cong \frac{C_{vib}}{R}$$

For this particular case of $HNNH_2^\ddagger$, the N-H bond is only ~ 55 kcal/mole while N-N is ~ 85 kcal/mole. Hence $k_H \gg k_r$ and $k_H \gg k_s[M]$ at flame temperatures and $k_{68} \sim k_f = 5 \times 10^{13}$ cm^3 $mole^{-1}$ s^{-1}, a typical recombination rate constant. Thus, this pathway to N-N bond formation is very rapid and this reaction plays a key role in the kinetics.

Summary

By combining laser diagnostics with improved computer algorithms for modeling laboratory flames, we have been able to

develop an improved mechanism for ammonia oxidation at high temperatures. The combination of absorption and laser induced fluorescence have yielded absolute concentrations of important flame
species. The sub-millimeter spatial resolution allowed us to
monitor species within the flame front, and the sub-ppm sensitivity allowed us to measure key reactive intermediates. These
concentration profiles provided us with the necessary data base
for development of a detailed mechanism. Since the algorithm
used accurately and efficiently accounted for the effect of
diffusive transport within the flame, we had the luxury of doing
a relatively simple, steady state experiment where we could
signal average to obtain adequate sensitivity for radicals. Since
transport was properly handled, we could focus exclusively upon
the kinetics. In this sense the modeling was similar to modeling
a shock wave experiment, but here we had the opportunity to
monitor reactive intermediates.

There are a host of high temperature systems which should now
be amenable to this combined diagnostic/modeling approach. It
should give kineticists an extra weapon with which to approach
complex systems.

Literature Cited

1. Morley, C., Eighteenth Symposium (International) on Combustion, Combustion Institute, Pittsburgh, PA (1981), p. 23, and references therein.
2. Lyon, R. K., U.S. Patent No. 3,900,554 (1975).
3. Dean, A. M.; Hardy, J. E.; Lyon, R. K.; Nineteenth Symposium (International) on Combustion, Combustion Institute, Pittsburgh, PA (182), p. 97.
4. Fujii, N.; Miyama, H.; Koshi, M.; Asaba, T.; Eighteenth Symposium (International) on Combustion, Combustion Institute, Pittsburgh, PA (1981), p. 873, and references therein.
5. Fisher, C. J., Combust. Flame, 30, 143 (1977), and references therein.
6. Chou, M.S.; Dean, A. M.; Stern, D.; J. Chem. Phys., 76, 5334 (1982).
7. Chou, M. S.; Dean, A. M.; Stern, D.; J. Chem. Phys., 78, 5962 (1983).
8. Dean, A. M.; Chou, M. S.; Stern, D.; Int. J. Chem. Kinetics (submitted).
9. Chase, M. W,, Jr.; Curnutt, J. L.; Downey, J. R., Jr.; McDonald, R. A.; Syverud, A. N.; Valenzuela, E. A.; J. Phys. Chem. Ref. Data, 11. 695 (1982).
10. Piper, L. G.; J. Chem. Phys., 70, 3417 (1979).
11. Smooke, M. D., "Solution of Burner Stabilized Pre-Mixed Laminar Flames by Boundary Value Methods," Sandia National Laboratories Report 81-8040 (1982).
12. Menon, P.G.; Michel, K. W.; J. Phys. Chem., 71, 3280 (1967).

13. Emanuel, G., "Table of the Kassel Integral," Aerospace Report No. TR-0200(4240-20)-5 (1969).
14. Benson, S.W., Thermochemical Kinetics, 2nd Edition, Wiley, New York (1976).
15. Roose, T. R., Ph.D. thesis, Stanford University (1981).
16. Westley, F., "Table of Recommended Rate Constants for Chemical Reactions Occurring in Combustion," NSRDS-NSB 67 (1980).
17. Dean, A. M.; Steiner, D. C.; J. Chem. Phys., 66, 598 (1977).
18. Cattolica, R. J.; Dean, A. M.; Smooke, M. D.; "A Hydrogen-Nitrous Oxide Flame Study," Sandia Report SAND 82-8776 (1982).
19. Baulch, D. L.; Drysdale, D. D.; Horne, D. G.; Evaluated Kinetic Data for High Temperature Reactions, Vol. 2, Butterworth, London (1973).
20. Cohen, N.; Westberg, K. R.; "Chemical Kinetic Data Sheets for High Temperature Chemical Reactions," Aerospace Report No. ATR-82(7888)-3.
21. Warnatz, J.; Eighteenth Symposium (International) on Combustion, Combustion Institute, Pittsburgh, PA (1981), p. 369.

RECEIVED November 10, 1983

Formation of NO and N_2 from NH_3 in Flames

RICHARD J. BLINT and CAMERON J. DASCH

Physics Department, General Motors Research Laboratories, Warren, MI 48090

Ammonia oxidation at high temperatures is an interesting
kinetic system of significant technological importance
for NO formation and destruction. Recent work has
largely been devoted to the moderate temperature (1300K)
region in which NH_3 can quantitatively destroy NO. This
report describes flame studies (1800-2800 K) of NH_3 oxidation in which NH_3 is a model compound for the conversion of fuel-bound nitrogen to NO and N_2. A detailed (42
reactions) reaction scheme has been constructed from
literature rate constants which predicts our measured
flame speeds, major species profiles, and NO levels in
ammonia-oxygen-diluent flames. These predictions
require that the radical pool size and the N_2/NO branching can be correctly described. This mechanism has been
further tested against our measurements of NO emissions
from CH_4-air flames doped with NH_3 and NO. While in NH_3
flames NH_2 is pivotal, the Fenimore-type "loading"
experiments indicate a more prominent role for N atoms
and the Zeldovitch reactions in hydrocarbon flames. In
part, this is a result of the higher radical concentrations in hydrocarbon flames. Full reaction mechanisms
will be necessary to predict the relative conversion of
fuel bound nitrogen to NO and N_2.

The interactive kinetics of NH_3 and NO at high temperatures have
received a great deal of attention in recent years. Much of this
attention originated from the work of Fenimore who investigated
the conversion of fuel bound nitrogen to NO in many flames. He
found two important properties of the NO production: 1) the yield
of NO was independent of the fuel-nitrogen type and 2) the NO
tended to be self limiting such that a saturation value of NO
could not be exceeded.

Fenimore ($\underline{1}$) formulated a schematic, two step kinetic

0097–6156/84/0249–0087$06.00/0

representation which described this saturation effect. The model
assumes that all the fuel nitrogen passes through a reactive,
amine intermediate N*. The first step is a general oxidation
reaction which forms NO

$$N^* + Ox \xrightarrow{k_1^*} NO + products$$

and the second step destroys both NO and the NO precursor

$$N^* + NO \xrightarrow{k_2^*} N_2 + products$$

The NO yield, [NO]/[fuel-N], is a function only of
[NO]sat$\equiv k_1^*$[Ox]/k_2^*. This formulation for the fuel-nitrogen conver-
sion to NO has been further tested by Haynes et al.(2,3) and by
Fenimore (4,5).

The essential feature of this formulation is the second reduction
step. This reaction had previously been recognized as important
from the fast rate of decay of NO produced in ammonia flames
(6,7). This reaction process has been exploited in the Exxon
DeNOx process for NO removal by the controlled addition of NH_3 to
combustion effluents.

The simple phenomenology and large economic importance for fuel
nitrogen conversion has encouraged many investigators to try to
identify the N* intermediate and its reaction rates k_1^* and k_2^*.
Neither Kaskan and Hughes (8) or Fenimore (3,4) could conclusively
demonstrate either NH_2, NH, or N as being exclusively pivotal.
Direct measurements of the NH_2 + NO reaction rate have shown it to
be fast (9,10) and important, especially at lower temperatures
such as 1300K where the DeNOx process is most efficient. This
lower temperature regime is well understood as a result of experiments and extensive kinetic modeling (11,12).

Our approach to the fuel nitrogen conversion to NO problem has
been to examine the kinetics of NH_3 as a model compound using
detailed flame calculations with modelable flame experiments.
This work has emphasized the major flame properties including the
flame speed, temperature and major species spatial profiles, and
the post-flame NO concentration. These measurements and calculations are performed on stationary free flames.

In this paper we first summarize the ability of the NH_3 kinetic
mechanism of Dasch and Blint (13) to describe the major properties
of ammonia-oxygen-diluent flames. This mechanism is constructed
solely from literature rate constants and is specifically valid
only for flames leaner than ϕ=1.2 Having validated the ammonia
mechanism, we then investigate the yield of NO from methane-air
flames which have been doped with NH_3 and NO. The NH_3 doped

methane flames are a typical model system for fuel-nitrogen NO production. The kinetic modeling(14) of this complete nitrogen and hydrocarbon flame system and its success represents a major advance in the full representation of the fuel-nitrogen problem. This work relies on the CH_4 mechanism of Warnatz(15) which has been tested against many flame properties including detailed temperature species profiles(16). There are no C-N coupling reactions between the CH_4 and NH_3 mechanisms although these reactions are known to be important for the production of "prompt NO" in rich flames ($\phi > 1.2$).

Miller et al. (17) have performed similar calculations for the ammonia flame experiments of Fenimore and Jones (6) and Maclean and Wagner (7). Dean and coworkers (this Symposium) have also performed experiments and calculations with an emphasis on radical intermediates.

These calculations show that in ammonia flames NH_2 is marginally more important than NH and N. Furthermore, the important oxidant changes with stoichiometry. Mixed CH_4-NH_3-NO flames show even a wider variation in the relative role of NH_2, NH, and N. In near stoichiometric NO doped methane flames, of course, the Zeldovitch mechanism involving N is exclusively important. Surprisingly, N is also most important in the NH_3 doped CH_4 flames. This is a consequence of the larger radical pool in hydrocarbon flames than NH_3 flames. Under both these extremes most of the doped nitrogen appears as NO. Under combined NH_3 and NO doping the flames are somewhat more similar to ammonia flames in which NH_2 is important for the reduction of NO to N_2. These combined NH_3-NO doping experiments exercise more rigorously the reductive parts of the mechanism than the Fenimore-type NH_3 doping experiments.

It appears that the quantitative prediction of Fenimore's simple [NO]sat parameter will require a large kinetic mechanism. The combined NH_3 and NO doping experiments test Fenimore's two step mechanism in a new way, but it is found that the single [NO]sat parameter can still correlate the results(14).

Experiments

The experimental methods are extensively described in earlier papers (13,14,16). Temperature and major species profiles were determined by spontaneous Raman spectroscopy in free flames stabilized on a slot burner. Flame speeds were determined by particle tracks on the same burner and by the Guoy method on conical flames. Total NOx measurements were performed in the postflame region of a flat, water cooled, Meker burner. On the flat burner gas flows were adjusted to give a slightly wrinkled flame corresponding to free flame conditions. Gas samples are extracted with a water cooled, quartz microprobe analyzed with a chemiluminescent

analyzer. Temperatures measured simultaneously with a corrected
thermocouple agree well with calculated adiabatic flame temperatures. The temperatures were constant for at least 5 cm above the
burner. The NOx concentrations typically varied less than 10%
within 1.0 cm of the flame front and then were constant for at
least 5 cm. The systematic uncertainly of the NOx measurements is
±10%.

Computational Method

The experimental observables for the free flames in this study
were modeled using the GMR flame program(18). This program solves
the unsteady state species and enthalpy conservation equations and
allows them to progress in time until the steady flame solution is
obtained. Boundary conditions for free flames were used. Three
types of information are required for these calculations: 1) the
reaction mechanism, 2) the transport properties and 3) the thermodynamic properties of all the species. The reaction mechanism
(given in Ref. 16) is constructed from literature values. The 10
H-O reactions are well established. The 32 additional amine and
NO reactions are largely taken from Ref. 19, 9, and others.
Although NH_i-NH_j hydrazine reactions were tested, they were not
important for these near stoichiometric flames and were not
retained in the mechanism. Selected values are taken for reactions which are radical branching or termination reactions and
hence especially important. The 65 additional reactions of the
CH_4 mechanism are from Warnatz(20). There are no C-N coupling
reactions between the NH_3 and CH_4 mechanisms, although it is known
that these are important in richer flames. The diffusion and con-
duction terms in the flame program are calculated from Stockmayer
potential parameters in the Fick's Law level of approximation
(21). The thermodynamics are taken from the JANAF compilation up
through the 1978 revision. The heat of formation for NH was taken
as 84.7 kcal/mole (22).

Pure NH_3 Flames

To test the reaction mechanism pure ammonia-oxygen-diluent flames
were studied under a wide range of stoichiometries and dilutions.
Peterson and Laurendeau(23) had previously studied NH_3 doped H_2
flames which elucidate much of the NH_i oxidation chemistry. The
present experiments probe the same reactions and also reactions
between nitrogeneous species. Flame speeds were first considered
in order to establish a subset of literature reaction rates which
would describe this major flame feature. Comparison of the other
flame features confirm that the flame speed is a good test bed for
the reaction mechanism.

The difficulty and the value of evaluating a reaction mechanism
based on flame speed over such a large range of dilutions and

equivalence ratios is that reaction effects at a particular temperature range or flame condition could be inadequate for those at
another. These flame speed data occur over the adiabatic flame
temperature range of 2270 to 2900 K.

Flame Speeds. Overall, the calculated flame speeds agree well
with experiment. Flame speed calculations as a function of diluent concentration with $\phi=0.8$ (the peak of the stoichiometric
dependence) are shown in Figure 1. The calculated flame speeds
are less temperature dependent than experiment. Considering the
stoichiometric dependence of the flame speeds, the calculations of
flame speed are in greater error away from the peak ($\phi=0.8$) and
exceed the experimental error at both rich and lean limits.

Flame Profiles Major species and temperature profiles provide a
detailed description of the structure of these flames. Three
flames at initial O_2 to N_2 ratios of 0.58 were probed using laser
Raman spectroscopy. the equivalence ratios were 0.7, 1.0 and
1.29. The calculated and measured temperature profiles for the
$\phi=0.7$ flame (given in Figure 2) shows good agreement. Similar
agreement is found at $\phi=1.0$ and 1.27. Figure 2 also gives the
oxidant decay profile which is another measure of the flame width.
Since the ratio of O_2 to N_2 is being measured, differences in the
calculated and the experimental species profiles are due to difference either in the decay rate of the oxygen or in the growth
rate of the nitrogen. For the flames measured, the O_2/N_2 ratio
tends to drop somewhat more quickly than the calculations. In the
leaner two flames the H_2 to N_2 profile reaches a maximum at about
1800K, well before the temperature maximum. The calculations
reproduce the spatial distribution and stoichiometric trends of
the H_2 profile but are a factor of 3-10 low in magnitude. While
the calculated flame speeds show deviations as a function of
equivalence ratio, the flame widths seem to be uniformly well
described.

Nitrogen Oxide Concentrations. The major objective of this work
is the understanding and prediction of the NO concentrations. The
equivalence ratio dependence of the NO is given in Figure 3. The
experimental and calculated values are also compared with the
equilibrium NO concentrations for the corresponding adiabatic
flame. As has been previously observed the NO decreases with
increasing equivalence ratio, exhibiting a slight maximum at 0.7.
as has been previously observed. As seen, the NO is always
"super- equilibrium" and more so in rich flames. Generally the
measured and calculated NO concentrations are the same except for
very rich flames. The calculations also reproduce the steep temperature dependence of the NO concentration. Not only does the NO
concentration increase with flame temperature, it is progressively
driven more "super-equilibrium".

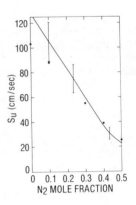

Figure 1.
Experimental and calculated
flame speeds vs. initial mole
fraction of N_2 in ammonia-oxygen-
nitrogen flames with $\phi = 0.8$.

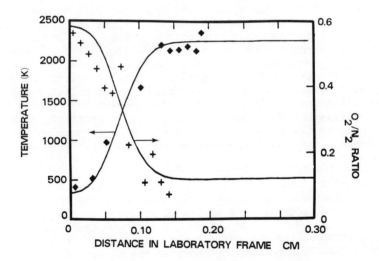

Figure 2. Experimental and calculated spatial profiles of temperature and O_2/N_2 ratio for a $\phi = 0.7$ ammonia-oxygen-nitrogen flame.

Major Pathways and Reactions Effects. As seen above the
experimental data (flame speeds, flame widths and NO concentrations) are well described by the kinetic mechanism. A schematic
of the whole kinetic chain for the combustion of the nitrogeneous
species is shown in Figure 4. The relative influence of reactions
on a species are determined by a flux analysis (13,15). This is a
flame averaged analysis, but it typically gives results similar to
a comparison of net reaction rates in the reaction zone. To further test the effect of the specific reaction changes we recalcu-
late selected flames.

One important conclusion from such analyses is that most intermediate species are in steady state or "kinetically limited". This
condition arises from very fast radical reactions. These lead to
tight kinetic coupling and chemical redistribution among the
intermediates much more quickly than changes due to advection or
diffusion. The consumption reactions for individual species will
vary the concentration to match its production rate. In the
flames studied these rates never differ by more than 20% for all
the radicals (H, O, OH, HO_2, HNO, NH_2, NH, N) comprising the "radical pool". The kinetically limited state allows some simplifying
relations between the intermediates to be expressed, and these
have been discussed by Blint and Dasch(13) for a number of the
nitrogenous intermediates. However, this state also implies that
the size of the intermediate pool will depend on the difference
between chain branching and terminating reactions. These reactions can be quite slow relative to the chain propagating reactions but have a larger influence since they control the radical
pool size. These reactions are separately considered following
the description of the primary pathways.

In these flames over 95% of the nitrogen atoms are combusted to
N_2. Since N_2 is predominantly formed by reactions of NO with NHi,
approximately half the nitrogen atoms are cycled through NO while
the other half are oxidized no further than NHi. Since the amine
radicals contribute to the N_2 formation in the rough ratios of
3:1:1 for NH_2:NH:N, the primary pathway to N_2 can be written by
the two sequences shown in heavy line. These sequences account
for about two thirds of the NO and N_2 formation. While this indicates a pivotal role for NH_2, neither NH nor N reactions with NO
can be eliminated from the reaction scheme.

One of the dominant reactions is NH_2 with NO branching to the two
sets of products,

$$NH_2 + NO \rightarrow N_2 + H + OH \qquad\qquad (R1)$$
$$\rightarrow N_2 + H_2O \qquad\qquad (R2)$$

As discussed in Ref. 16 the branching ratio of 70% R1 and 30% R2

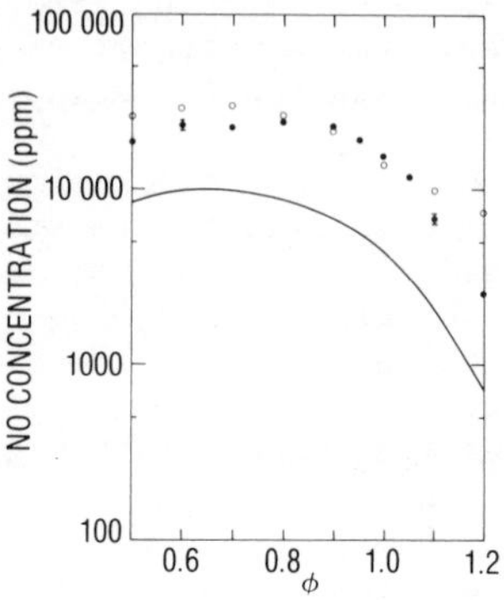

Figure 3. Experimental and calculated maximum NO concentrations as a function of equivalence ratio for ammonia-oxygen-nitrogen flames with fixed initial mole fraction of N_2 = 0.4.

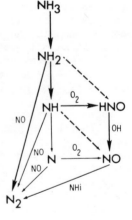

Figure 4. Schematic of the detailed kinetic mechanism for ammonia combustion.

was used. To test the effect of changing the branching ratio the
flame speeds were recalculated using the branching ratio of 40/60
suggested by Silver, et. al.(9). For flames with initial N_2 dilutions of 0.4 the flame speeds are slower and the dependence as a
function of equivalence ratio is modified. The maximum in the
flame speed is shifted richer than ϕ=0.8. The flux analysis for
reactions R1 and 2 shows a reduction of about 30% for the flux of
the products. The flame width of the ϕ=0.5 N_2(initial)=0.4 flame
increases by almost a factor of three with the change in the
branching ratio. Clearly this branching ratio has a large effect
on the flame properties of ammonia flames.

Another reaction with a significant effect on the flame properties
is

$$NH + O_2 \rightarrow HNO + O$$

Little is known about this reaction. We have used the value suggested by Peterson and Laurendeau(23). The sensitivity was tested
with the rate constant that Branch et. al. (12) developed for the
lower temperatures of the "Thermal Denox" problem. Using this
smaller rate constant (at flame temperatures) on two flames (ϕ=0.5
N_2(initial)=0.4 and ϕ=0.8 N_2(initial)=0.1) we find 15-20%
decreases in the flame speed and 40-50% decreases in the flux contribution from this reaction. Overall the effect of reducing the
$NH + O_2$ rate is to reduce the rate of production of HNO in each of
the flames; no significant variation of this effect with equivalence ratio was observed. Changing the products of the reaction
to NO and OH as suggested by Binkley and Melius(24) gives results
similar to reducing the rate. With these products fewer radicals
are generated; hence the flame speed is reduced by more than 20%
and the flame widths thicken slightly.

NO Yield in Pure and Doped CH_4 Flames

Having validated the mechanism on ammonia-oxygen flames, the yield
of NO from nitrogen doped CH_4-air flames was examined. Both NH_3
and NO doping were investigated. Only post-flame NO concentrations were measured. These are compared with calculations of the
full kinetics and with adiabatic equilibrium calculations. The
calculated profiles indicate the complexity of the NO dynamics in
these flames. The temperature and major species profiles in the
undoped flames had been studied in earlier work(16). Three near
stoichiometric methane-air flames having initial equivalence
ratios(ϕ) of 0.8, 1.0 and 1.2 are diluted with less than 5 volume
percents of NH_3 or NO. In this section NO concentration is
expressed both as a mole fraction and as a fraction of the total
nitrogen concentration:

$$NO\ fraction \equiv [NO]/([NO]+2[N_2]+[NH_3]+[NH_2]+[NH]+[N]+ [HNO]+2[N_2O])$$

This approach removes any effects due to changing numbers of molar
species within the flame.

Pure Methane Flames. In a pure methane flames the NO fraction
undergoes a small jump passing through the flame front, then
increases linearly with time-distance. This linear increase is
due to low NO concentration after the flame front (approximately
30 ppm in a stoichiometric flame) which is far less than the equilibrium concentration of NO (3100 ppm). The reaction(R3) driving
the formation of NO is part of the Zeldovitch mechanism:

$$O + N_2 \rightarrow NO + N \qquad (R3)$$
$$N + O_2 \rightarrow NO + O \qquad (R4)$$
$$N + OH \rightarrow NO + H \qquad (R5)$$

The reverse reactions are negligible. The Zeldovitch mechanism is
rate limiting in these near stoichiometric flames having no added
nitrogen species. There is no significant NH or N production.

NO doped Methane Flames. Most of the NO passes through the flame
unreacted, and the NO primarily acts as an inert diluent. Dilut-
ing the flames with 1.5% NO causes the adiabatic flame temperatures to drop by about 100 K and the calculated flame speeds to
decrease by about 10%. The NO fraction decreases by less than 10%
through the flame front for these flames which were doped far
above the equilibrium NO concentrations (420-3700 ppm depending on
stoichiometry). In the experiments there was no post flame decay
in the NO concencentrations (±5%), while the calculations do show
some decay (<2%/mm) which depends upon stoichiometry. The calculated post flame conversion rate of NO to N_2 decreases as the
super-equilibrium radicals (e.g.; O and OH) decay.

The very slow reduction of the NO through the flame is a result of
the back reactions of the extended Zeldovitch mechanism. N atoms
are slowly produced from NO by reactions R4 and R5 but then rapidly react again with NO via R3. R3 is the major reaction producing N_2 with its fastest rate occuring at the radical maximum just
past the inflection point in the temperature profile.

Two minor reactions do produce species which directly involve the
full ammonia reaction mechanism. These produce NH_2 and NH from
HNO. N atom production from NH contributes only of the order of
5% of the total N atom production. Even this small contribution
of the $HNO \rightarrow (NH_2, NH) \rightarrow N$ pathway is restricted to the high radical
region of the flame front.

NH_3 Doped Methane Flames. While the addition of NO to a methane-
air flame acts primarily as a diluent, the addition of NH_3 richens
the mixture leading to competition for the available oxygen. This
richening effect also reduces the equilibrium NO concentration of
these near stoichiometric flames ($0.8 < \phi < 1.2$).

The measured post-combustion NO concentrations values are generally similar to the calculated concentrations(Figure 5). The measured NO concentrations tend to be higher at the higher ammonia
dopings. While the NO concentrations increase with NH_3 doping,
the fraction of ammonia converted into NO decreases with a corresponding increase in the N_2 fraction. This is the saturation
behavior extensively studied by Fenimore (1,4) and Haynes(2,3).
After the NO reaches a critical steady-state value, additional
fuel-nitrogen is converted to N_2. This behavior is also seen in
the calculated NO profiles.

Interestingly, the major source reaction of N_2 in these ammonia
diluted methane-air flames is again reaction R3. The "H shuffle"
reactions form a fast pathway from NH_3 to N atoms. The extended
Zeldovitch reactions R3, R4, and R5 then determine the NO and N_2
production. This behavior is in marked contrast to the pure ammonia flame pathways (Figure 4) where NH_2 and NH were significant N_2
soucrces, and HNO was the major precursor of NO.

This difference in behavior between NH_3 and CH_4 flames results
from an approximate doubling of the radical concentration from a
$\phi=1.0$ ammonia flame to a $\phi=1.0$ CH_4/air flame. This change pro-
motes the NH→N pathway over the NH→HNO pathway.

Combined NO and NH₃ Doped Methane Flames. Adding both NO and NH_3
to methane-air flames provides the opportunity to enhance the
reducing effect of the NHi + NO reactions. Typically the flames
are doped with NO concentrations far in excess of the equilibrium
concentrations; consequently we see NO concentrations which are
appreciably reduced by the NH_3 addition.

The calculated NO concentrations for the low dopings of NH_3 show
good agreement with experiment. Table I is a comparison of the

Table I.Experimental, calculated and equilibrium NO concentrations
in ppm for the methane-air flames diluted with 1.5% NO.

ADDED NH₃(%)	$\phi = 0.8$			$\phi = 1.0$			$\phi = 1.2$		
	exp.	cal.	eq.	exp.	cal.	eq.	exp.	cal.	eq.
0.0	13500	12000	3700	12200	12500	3100	12500	12700	420
0.3	**	**	**	**	**	**	11000	11700	350
1.0	14200	12800	3800	13500	13300	2400	8000	9600	232
3.0	14200	11500	3600	13800	9300	914	7800	5000	70
5.0	14900	10200	2700	12200	6400	250	**	**	**

**flame not calculated

measured, calculated and equilibrium NO concentrations for the flames diluted by 1.5% NO and doped with varying amounts of NH_3. above equilibrium. Deviations seem to occur primarily at high dopings of ammonia. Typically the calculations give low values. The agreement is somewhat worse than the table indicates since the calculated NO concentrations slowly decrease with increasing distance post-flame while the experimental concentrations are constant.

Figure 6 shows calculated spatial profiles of NO fraction and temperature in a $\phi=0.8$ methane-air flame doped with 5% NH_3 and 1.5% NO. This is a typical plot of the NO fraction for the combined loading experiments. There is an initial precipitous drop in the NO concentration early on in the temperature profile, followed by a slight rise and then the very slow drop in the NO concentration with distance. The intermediate NO peak shown in this particular example is frequently found, although some NO profiles show only the initial rapid drop and the very slow secondary reduction. The initial NO destruction found in these high NO dilutions (1.5%) is due to the NH_2+NO pathway. In contrast to the NH_3 doped flames, these flames do have NO early in the temperature rise where the NH_2 occurs. The reduction of the NO as shown in Figure 6 occurs in a very narrow temperature range. This initial NO destruction is a major portion of the NO dynamics over the entire flame. Specifically, R1 is found to account for at least 1/3 of the NO converted into N_2 as measured by a flux analysis of N_2. The midflame features are due to the extended Zeldovitch reactions and show a complicated interplay of temperature and NH_3 decay. Comparing the NH_3 and NO concentrations in this zone, the NO rise is clearly from NH_3 which did not participate in the early reduction of NO.

These mixed NO and NH_3 experiments show that the full temperature range and all the branches of the reaction mechanism are important for describing the evolution of fuel-nitrogen even in near stoichiometric flames.

Conclusions

Pure ammonia flames are studied experimentally and theoretically to isolate the amine chemistry. Ammonia and NO doped methane-air flames are also studied as models for flames with fuel-bound nitrogen and flames diluted with NO from EGR. We conclude from these studies:

1. Pure ammonia flames follow major pathways which are different from doped hydrocarbon flames. NH_2 is pivotal for determining relative N_2 and NO yields in ammonia flames.

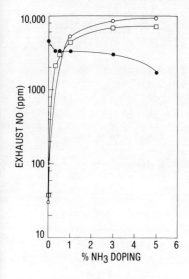

Figure 5.   Calculated and measured maximum NO concentrations for ϕ = 0.8 methane-air flames as a function of total NH_3 doping.

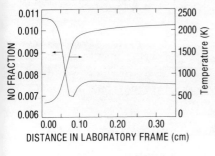

Figure 6.   Temperature and NO fraction spatial profiles for ϕ = 0.8 methane-air flame doped with 5% (vol) NH_3 and 1.5% NO.

2. NO alone doped into methane flames is only reduced by the slow
   three-reaction Zeldovitch mechanism. No appreciable reduction
   due to the ammonia reaction mechanism is found.

3. Contrary to pure ammonia combustion where the primary source of
   N_2 comes from reactions of NH_2 with NO, ammonia doped into
   methane-air flames is combusted to N_2 via the N + NO reaction.

4. Mixtures of NO and NH_3 in methane-air flames exhibit separate
   regions of reducing NO to N_2 and then oxidizing NH_3 to NO.
   The reducing region is due to the NH_2+NO reactions, while the
   oxidizing zone is dominated by the Zeldovitch reactions. The
   relative importance of the two mechanisms is dependent on the
   shape of the temperature profile of the flame. It is
   concluded that the reducing effect of fuel-nitrogen on
   hydrocarbon fuel mixtures containing NO is dependent on the
   flame structure.

Acknowledgments

We wish to thank R.E. Teets and J.H. Bechtel for helpful discussions, L. Green and S. Howell for technical assistance, and A.D.
Gara for encouragement and support.

Literature Cited

1. Fenimore, C.P. Combust. Flame 1972, 19, 289.

2. Haynes, B.S.; Iverach, D.; Kirov, N.Y. Fifteenth Symposium
    (International) on Combustion 1975, p. 1103.

3. Haynes, B.S. Combust. Flame 1977, 28, 81.

4. Fenimore, C.P. Combust. Flame 1976, 26, 249.

5. Fenimore, C.P. Seventeenth Symposium (Internation) on
    Combustion 1979, p. 661.

6. Fenimore, C.P; Jones, G.W. J. Phys. Chem. 1961, 65, 298.

7. MacLean, D.I.; Wagner, H. Gg. Eleventh Symposium
    (International) on Combustion 1967, p. 871.

8. Kaskan W.E.; Hughes, D.E. Combust. Flame 1973, 20, 381.

9. Silver, J.A.; Kolb, C.E. J. Phys. Chem. 1982, 86, 3240.

10. Roose, T.R.; Hanson, R.K.; Kruger, C.H. Eighteenth
    Symposium(International) on Combustion 1981, p. 853.

11. Lyon, R.K.; Dean, A.M.; and Hardy, J.E. Nineteenth Symposium (Internation) on Combustion 1983.

12. Branch, M.C.; Kee, R.J; Miller, J.A. Combust. Sci. Tech. 1982, 29,147.

13. Dasch, C.J.; Blint, R.B. "A Mechanistic and Experimental Study of Ammonia Flames", General Motors Research Laboratories Report GMR-4232, and Eastern States Fall Meeting, The Combustion Institute, 1982, Paper ESS/CI 82-71.

14. Blint, R.B.; Dasch, C.J. "Experiments and Calculations of NO Yield from NH_3 and NO in CH_4 Flames", General Motors Research Research Report, 1983.

15. Warnatz, J. Ber. Bunsenges. Phys. Chem. 1979, 83,950.

16. Bechtel, J.H.; Blint, R.J.; Dasch, C.J.; Weinberger, D.A. Combust. Flame 1981, 42, 197.

17. Miller, J.A.; Smooke, M.D.; Green, R.M.; Kee., R.J. "Kinetic Modeling of the Oxidation of Ammonia in Flames", Western States Section Fall Meeting, The Combustion Institute, 1982, Paper WSS/CI 82-93.

18. Dasch, C.J.; Blint, R.B. "An Improved Spalding-Stephenson Transformation for One Dimensional Flame Calculations", accepted for publication in Combust. Sci. Tech.; and Western States Section Fall Meeting, The Combustion Institute, Paper WSS/CI 82-89, 1982.

19. Salimian, S.; Hanson, R.K. Combust. Sci. Technology 1980, 23, 225.

20. Warnatz, J. Eighteenth Symposium (International) on Combustion 1981, p.853.

21. Hirschfelder, J.O.; Curtiss, C.F.; Bird, R.B. Molecular Theory of Gases and Liquids John Wiley & Sons Inc., New York, 1954.

22. Piper, L.G. J. Chem. Phys. 1979, 70, 3417.

23. Peterson, R.C.; Laurendeau, N.M. Central States Section, The Combustion Institue, 1982, Paper CSS/CI 82-15.

24. Binkley, J.S.; Melius, C.F. "Oxidation of the NH and NH_2 Radicals", Western States Section Fall Meeting, The Combustion Institute, 1982, Paper WSS/CI 82-96.

RECEIVED November 10, 1983

Reactions of NH and NH₂ with O and O₂

Theoretical Studies

C. F. MELIUS and J. S. BINKLEY

Sandia National Laboratories, Livermore, CA 94550

Calculations involving fourth-order Møller-Plesset
perturbation theory with bond additivity corrections
were used to investigate the reactions of NH_2 and
NH with O and O_2. New heats of formation were used
for NH_2 (47 kcal-mole^{-1}) and NH (87 kcal-mole^{-1}). We
find that the NH_2O_2 complex is barely stable. Thus, the
rate constant for $NH_2 + O_2$ should be very small with the
most probable products at high temperatures being NH_2O+O.
For $NH+O_2$, we find that the reaction forming HNO+O is
spin forbidden on the lower lying singlet surface. Thus,
at room temperature the likely products are NO+OH and
$H+NO_2$, which are accessible on the singlet surface.
     For NH_2+O and NH+O, we find that stable inter-
mediate complexes can be formed with no activation
barrier. These can undergo direct dissociation by
losing a hydrogen or can undergo a 1,2-hydrogen shift
with further dissociation forming OH. The reaction pathway involving molecular complex formation should dominate
at room temperature while direct hydrogen abstraction by O
atoms should dominate at high temperatures.

The reactions of NH_2 and NH with O_2 and O play important roles in
the oxidation of ammonia and fuel bound nitrogens as well as in
the reduction of NO_x's(1-12). However, the rate constants for
these reactions differ drastically in the various chemical
reaction models which have been developed (1-6). Even the
experimentally measured rate constants show a wide variation and
differ significantly from the corresponding isoelectronic counter-

0097-6156/84/0249-0103$06.00/0
© 1984 American Chemical Society

parts involving carbon and oxygen (6,13-23). Furthermore, it is
not even clear what the products of these reactions should be.

To aid the modelers in developing improved reaction mechanisms as well as to aid the experimentalists in their interpretation of the data, we have calculated the energetics of molecular
intermediates and products arising from these reactions. Our
approach was to use the highly accurate fourth order Møller-
Plesset perturbation theory (24) with bond-additivity corrections.(25)

In the next section, we briefly present the theoretical
approach. We then present and discuss the results for the
reactions NH_2+O_2, $NH+O_2$, NH_2+O, and $NH+O$.

Theoretical Approach

The method is divided into two distinct steps. The first step
consists of ab initio computation of the total energies of the
desired molecule. This is accomplished by first determining the
theoretical equilibrium geometry of the molecule using the
6-31G*(26) basis and spin-restricted Hartree-Fock (27) theory
for closed-shell molecules and spin-unrestricted Hartree-Fock
(28) theory for open-shell molecules. Using these geometries
two additional calculations are performed. Zero-point vibrational
corrections are derived using analytically computed (29) second-
derivatives of the energy with respect to the internal degrees of
freedom. In the second calculation electron correlation effects
are determined using Møller-Plesset perturbation theory carried
out to fourth-order (24) using the 6-31G** (26) basis
(denoted MP4/6-31G**). Combining the data from these two calculations leads to a vibrationally corrected, electron-correlated
total energy for a given molecule.

These theoretical calculations are subject to systematic
errors arising from basis set truncation and neglect of higher
perturbation order electron correlation effects. Therefore, we
have added a second step to our procedure in which we attempt to
account for these deficiencies by adding empirically derived
Bond-Additivity-Correction (BAC) factors that are based on the
types of bonds present in the given molecule. The correction
factors for NH (9.4 kcal-mole^{-1}) and OH (10.7 kcal-mole^{-1}) bonds
are obtained as the difference between the theoretical and

experimental energies for complete dissociation of NH_3 and OH_2, respectively, averaged over the number of bonds in the molecule. In a similar fashion, correction factors for the nitrogen-oxygen and oxygen-oxygen bond types are developed from a comparison of the theoretical results with various experimental values of N-O-H containing molecular species. In this case the bond additivity corrections depend both on the bond distance as well as the bond type. An additional correction factor for spin contamination in the spin-unrestricted Hartree-Fock wavefunction is included.

Using the theoretically computed total energies, ΔH_f^0 values were be obtained for the appropriate H, N, O-containing species. This procedure tacitly assumes the transferability of the corrective factors from one molecule to another and is subject to any limitations imposed by the original experimental values. The full details and a critical anlysis of this procedure will be published separately.

Results and Discussion

The calculated heats of formation for various stable and meta-stable intermediates and products for the reactions NH_2+O_2, $NH+O_2$, NH_2+O, and $NH+O$ are shown in Figures 1-4. The heats of formation for the polyatomics agree with known experimental heats of formation to within several kcal-mole^{-1}. The exceptions are NH and NH_2 where the experimental estimates vary widely. (Experimental values for NH_2 range from 40-46 kcal-mole^{-1} while experimental values for NH range from 81-90 kcal-mole^{-1}). We therefore have used our recently calculated values of $\Delta H_f^0(NH_2)$ = 47 kcal-mole^{-1} and $\Delta H_f^0(NH)$ = 87 kcal-mole^{-1} as being the most reliable (25).

Included in the figures are schematic curves which connect various isomers with each other and with reactants or products. Transition state potential energy barriers have <u>not</u> been calculated for these processes. However, we have included these schematic curves (with an indication of whether the transition state barrier should be large, small, or nonexistent) in order to help the reader follow possible reaction mechanisms.

NH$_2$+O$_2$. Several reaction schemes have been proposed for NH_2+O_2 (ΔH values for all reactions were computed in this study):

$$NH_2+O_2 \longrightarrow NO+H_2O \qquad \Delta H = -83 \text{ kcal-mole}^{-1} \qquad (1)$$

proposed by Gesser (7),

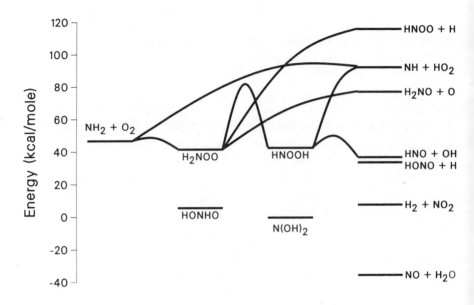

Figure 1.   Calculated energies of possible intermediates and products for the reaction of $NH_2 + O_2$. Vertical scale represents ΔH_f^0 at OK. Connecting curves schematically indicate possible reaction pathways.

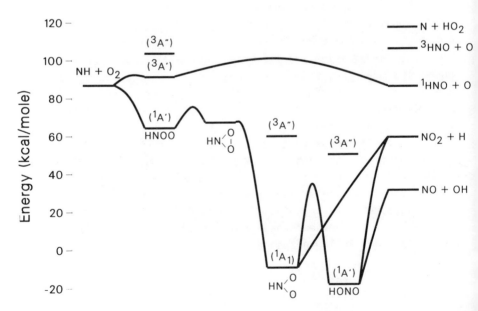

Figure 2.   Same as in Figure 1 for $NH + O_2$.

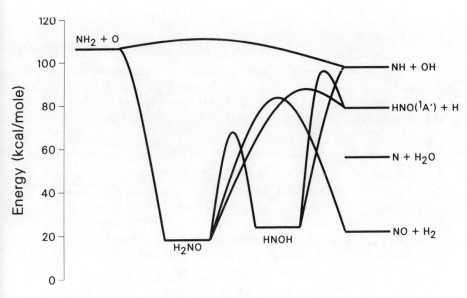

Figure 3. Same as in Figure 1 for NH_2 + O.

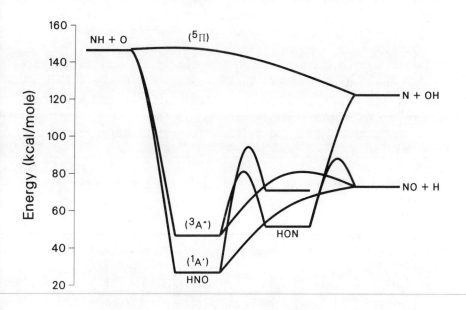

Figure 4. Same as in Figure 1 for NH + O.

$$NH_2+O_2 \longrightarrow HNO+OH \quad \Delta H = -10 \text{ kcal-mole}^{-1} \quad (2)$$

first proposed by Husain and Norrish (8),

$$NH_2+O_2 \longrightarrow NH_2OO \quad \Delta H = -5 \text{ kcal-mole}^{-1} \quad (3)$$

first proposed by Jayanty, et al. (13) and favored by Hack, et al. (14), and

$$NH_2+O_2 \longrightarrow NH+HO_2 \quad \Delta H = +45 \text{ kcal-mole}^{-1} \quad (4)$$

first proposed by Takeyama and Miyama (9).

As can be seen from Figure 1, NH_2OO does not form a very stable complex, having a NH_2-O_2 bond energy of less than 6 kcal-mole^{-1}. Our calculations are consistent with the estimates by Pagsberg, et al. (15) and Benson (16) and imply that the interpretation of the experimental rate data by Hack, et al. (14) should be reexamined with respect to forming products, since the 1,3-hydrogen shift to form HNOOH should have a significant barrier. Thus, it is unlikely that reactions (1) and (2) as well as (4) via the HNOOH complex will have large rate constants. The recent theoretical calculations of Pouchan and Chaillet (17), using a less sophisticated subset of the method employed here, find NH_2OO to be unbound by 13 kcal-mole^{-1}. However, their conclusions are consistent with ours in tending to rule out reaction (3). Thus, while there are several exothermic products for the reaction of NH_2+O_2 (e.g., HONO+H, HNO+OH, NO_2+H_2, and $NO+H_2O$), there are no simple reaction pathways to form them.

Based on Figure 1, we believe the most probable reaction to be

$$NH_2+O_2 \longrightarrow H_2NO+O \quad \Delta H = +30 \text{ kcal-mole}^{-1} \quad (5)$$

as suggested by Benson (16), particularly at the higher temperatures in flame conditions. Such a mechanism is consistent with the observation that at room temperature the measured rate constant for NH_2+O_2 is negligibly small ($<6 \times 10^5$ cc mole^{-1} sec^{-1} determined by Lesclaux (18), $<5 \times 10^9$ cc mole^{-1} sec^{-1} determined by Pagsberg (15), $<9 \times 10^6$ cc mole^{-1} sec^{-1} determined by Cheskis et al. (21).

($\underline{19}$)). The lack of a stable complex explains why the room temperature rate is much smaller than that of the isoelectronic
reaction of CH_3+O_2 to form the CH_3OO complex ($\underline{14}$), which is
stable by 26 kcal-mole^{-1}. The endothermicity of 30 kcal-mole^{-1}
for reaction (5) (no barrier should exist on the exit channel) is
consistent with both the temperature dependence of 37.7 ± 2.6
kcal-mole^{-1} fit to induction times of various shock tube
experiments ($\underline{10}$) and with the need for a large activation energy
for the reaction of NH_2+O_2 used in various models of ammonia
oxidation ($\underline{1-3,5,6}$). The direct hydrogen abstraction reaction
(4) can also occur (on the quartet potential energy surface),
and should have a slight activation barrier in addition to being
endothermic.

$\underline{NH+O_2}$. Two reaction pathways have been proposed for $NH+O_2$.
One, suggested by Husain and Norrish ($\underline{8}$), involves a rearrangement
collision to form NO, i.e.,

$$NH+O_2 \longrightarrow NO+OH \qquad \Delta H = -56 \text{ kcal mole}^{-1} \qquad (6)$$

This reaction pathway is favored by Takeyama and Miyama ($\underline{9}$),
supported by Drummond ($\underline{12}$) and others ($\underline{6}$), and used by
Duxbury and Pratt ($\underline{3}$) and by Azuhata, et al. ($\underline{4}$) in their
modeling. The second reaction pathway involves the more direct
displacement reaction forming HNO, i.e.,

$$NH+O_2 \longrightarrow HNO+O \qquad \Delta H = -1 \text{ kcal mole}^{-1} \qquad (7)$$

While there has been some concern that reaction (7) may be
endothermic, our calculated heat of formation for NH of 87
kcal-mole^{-1} implies that (7) is essentially thermal neutral.
Reaction (7) is favored by Zetzsch and Hansen ($\underline{20}$) as being
simpler than (6) and is used by Fujii, et al. ($\underline{6}$), Miller,
et al. ($\underline{1}$), and Lyon, et al. ($\underline{2}$) in their modeling efforts.
In either case, the activation barrier used by the modelers is
small (3 kcal-mole^{-1}) or absent. The rate constant for $NH+O_2$
products at room temperature is small (5.1×10^9 cc mole^{-1}sec^{-1}
as measured by Zetzsch and Hansen ($\underline{20}$);$<2\times10^{10}$ cc mole^{-1}sec^{-1}
as measured by Pagsberg, et al. ($\underline{15}$)), although a rate constant of
3.6×10^{11} cc mole^{-1}sec^{-1} was measured indirectly by McConnell

To assist in distinguishing between the possible reaction pathways and rate constants, we present the calculated energetics of the $NH+O_2$ system in Figure 2. The system is complicated in that three potential energy surfaces must be considered. We can immediately rule out the quintet surface since no low-lying intermediate states or final products exist for this surface. The lowest reaction pathway exists for the singlet surface, for which the HNOO intermediate (stabilized by the N-O bond energy which is estimated to be 22-29 kcal-mole^{-1}) can be formed without any significant barrier. The HNOO species is isoelectronic with ozone, a biradical singlet species (28-30). The HNOO species can undergo ring closure followed by scission of the O-O bond to form the stable nitro hydride isomer. Our calculations indicate that the activation energy for this ring closing and reopening should be less than the 22-29 kcal-mole^{-1} bond energy formed from the incoming $NH+O_2$.

Having formed nitro hydride, there is sufficient internal energy to break the N-H bond to form $H+NO_2$, i.e.,

$$NH+O_2 \longrightarrow H+NO_2 \qquad \Delta H = -28 \text{ kcal mole}^{-1} \qquad (8)$$

or to undergo a 1,2-hydrogen shift to form the more stable HONO which can decompose to form NO+OH (reaction (6)) as well as $H+NO_2$ (reaction (8)). Under the appropriate conditions one could form the HNOO complex rather than the more stable HONO. It should be noted that the direct displacement reaction forming $HNO+O(^3P)$ (reaction (7)) cannot occur through the stable $HNOO(^1A')$ complex but must occur on the higher lying triplet surface. Removing the O atom from $HNOO(^1A')$ would produce $HNO(^3A)+O(^3P)$ analogous to the isoelectronic reaction $O(^3P)+O_2(^3\Sigma_g^-) \longrightarrow O_2(^3\Sigma_g^-) + O(^3P)$. However, unlike $O + O_2$, this reaction is endothermic by 19 kcal-mole^{-1}, which would explain why the rate constant for $NH+O_2$ is smaller than that for the isoelectronic $O+O_2$ (29) exchange reaction.

Reaction (7) will occur on the triplet surface through the metastable $HNOO(^3A')$ intermediate. This intermediate is calculated to be only slightly endothermic (4 kcal-mole^{-1}). There should be a very small activation barrier on the incoming reaction path. Of greater significance is the outgoing reaction path $(HNOO(^3A') \longrightarrow HNO+O(^3P))$ which must undergo a two-electron avoided curve crossing via a breaking of planar symmetry, giving rise to an additional 5-15 kcal-mole^{-1} energy barrier. Thus, the activation energy for reaction (7) should be larger than the 3 kcal-mole^{-1} used by the modelers.

Taking both surfaces into account, we conclude that the room temperature process should be dominated by reactions (6) and (8) on the singlet surface while at high temperatures, reaction (7) should dominate.

$\underline{NH_2+O}$. Two sets of products have been proposed ($\underline{22,23}$) for the reaction of NH_2 with atomic oxygen:

$$NH_2+O \longrightarrow HNO+H \qquad \Delta H = -28 \text{ kcal-mole}^{-1} \qquad (9)$$

and

$$NH_2+O \longrightarrow OH+NH \qquad \Delta H = -9 \text{ kcal-mole}^{-1} \qquad (10)$$

There has been some question as to whether or not reaction (10) is endothermic or exothermic. Based on our calculated heats of formation of NH and NH_2, reaction (10) should be exothermic by 9 kcal-mole^{-1}. The rate constant for NH_2+O at room temperature is 2.1×10^{12} cc mole^{-1}sec^{-1} as measured by Gehring, et al. ($\underline{22}$). This is smaller than the measured rate constants for the isoelectronic species CH_3 and OH reacting with O atom.

Our theoretical results (shown in Figure 3) indicate that both reactions (9) and (10) are possible. On the doublet surface, NH_2+O should form the H_2NO intermediate without an activation barrier. This complex can readily dissociate to form HNO+H (Reaction (9)). The radical intermediate is sufficiently exothermic that it can also undergo a 1,2-hydrogen shift to form HNOH which can dissociate to form either HNO+H or NH+OH (Reactions (8) or (9)).

There should also be sufficient energy for H_2NO to dissociate to $NO+H_2$,

$$NH_2+O \longrightarrow NO+H_2 \qquad \Delta H = -85 \text{ kcal-mole}^{-1} \qquad (11)$$

However, the dominant reaction is expected to be the direct displacement reaction (9) with smaller branching ratios for reactions (10) and even less for (11).

On the quartet surface, no intermediate complex exists that
is stable with respect to NH_2+O . However, hydrogen abstraction
(reaction (10)) can occur, having a small activation barrier.
On the other hand, there is no pathway to the exothermic products
$N(^4S) + H_2O$.

Thus, we conclude that at room temperature, the reaction
NH_2+O should be dominated by reaction (9) with some smaller
branching probability for reactions (10) and (11), all occurring
on the doublet surface with no activation energy. At higher
temperatures, the hydrogen abstraction reaction (10) should
dominate. We find no reason to expect the NH_2+O reaction rate
at room temperature to be appreciably smaller than its isoelectronic counterparts, CH_3+O and $OH+O$.

NH+O. The reaction NH+O is, in many respects, analogous to the
reaction of NH_2+O. Two sets of reaction products have been used in
combustion modeling, (1,3,5,) i.e.,

$$NH+O \longrightarrow NO+H \qquad\qquad \Delta H = -74 \text{ kcal-mole}^{-1} \qquad\qquad (12)$$
and
$$NH+O \longrightarrow OH+N \qquad\qquad \Delta H = -24 \text{ kcal-mole}^{-1} \qquad\qquad (13)$$

For reaction (12) an activation energy of 5 kcal-mole^{-1} is
used while for reaction (13) the activation energy used is very
small (0.1 kcal-mole^{-1}). In Figure 4, we present the results
for the isomeric complexes on the lowest singlet ($^1A'$) and
triplet ($^3A''$) surfaces. These results are consistent with
other work on the HNO system (34,35).

The HNO radical intermediates can be formed from NH+O without
an energy barrier. The formation energy of the complexes is
sufficient to either dissociate to H+NO (reaction (12)) or allow
the 1,2-hydrogen shift and dissociation to OH+N (reaction (13)).
The $^3A''$ surface is directly analogous to the doublet surface for
NH_2+O (Figure 3). For the $^3A''$ surface the dominant product should
be NO+H though some N+OH can be formed. On the other hand, HNO
formed on the $^1A'$ surface can only dissociate to H+NO since
formation of OH+N is spin forbidden. At higher temperatures, the
direct hydrogen abstraction reaction can occur, particularly on
the quintet surface, requiring a small activation energy.

Thus, for NH+O we conclude that the reaction rate should be

fast, with no activation energy for reactions (12) and (13) going through the HNO intermediate and with a slight activation energy for the direct hydrogen abstraction reaction (13).

Conclusions

The calculational approach used here involving fourth-order Møller-Plesset perturbation theory with bond additivity corrections (BAC-MP4), has been shown to be a powerful tool for determining heats of formation of molecular species and for analyzing possible reaction pathways. For NH_2 and NH reacting with O and O_2 we have been able to distinguish between likely and unlikely reaction products based on the stability of various reaction intermediates compared to the reactants and products.

The reactions studied here are typical of many systems in that they are complex, involving multiple potential energy surfaces with competing pathways, even when the same product channel exists. Thus, theoretical results such as these can be helpful both in analyzing experimental rate constants as well as in computer modeling of reaction systems. Additional work is needed to address the reaction pathways for products going to other products, since other parts of the potential energy surface may be involved. We are currently in the process of developing an approach for calculating transition state energy barriers (which were only included schematically in this paper), which can be of comparable accuracy to the results presented here.

Acknowledgments

The authors would like to give particular thanks to Dr. R. J. Blint of General Motors for encouraging our investigation of these N-H-O reactions and providing enlightening discussions involving the possible reaction mechanisms. The authors would also like to thank Drs. J. A. Miller, W. A. Goddard, and M. J. Frisch for helpful discussions involving various aspects of this research. This work was supported by the U. S. Department of Energy, Basic Energy Science, Chemical Physics Program.

Literature Cited

1. J. A. Miller, Combustion and Flame , 1981, 43, 81.
2. R. K. Lyon, A. M. Dean, and J. E. Hardy, Nineteenth Symposium
    (International) on Combustion, The Combustion Institute,
    Pittsburgh, PA, 1982, 97.
3. J. Duxbury and N. H. Pratt, 15th Symposium (Int.) Combust.
    (Proc.) 15th, 1975, 843.
4. S. Azuhata, R. Kaji, H. Akimoto, and Y. Hishinuma, 18th Symp.
    (Intl.) Combustion Proc. , 18th, 1981, 845.
5. C. J. Dasch and R. J. Blint, General Motors Research
    Publication-4232(1982).
6. N. Fujii, H. Miyama, M. Koshi, and T. Asaba, 18th Symp.
    (Intl.) Combustion Proc., 18th, 873 (1981).
7. H. Gesser, J. Am. Chem. Soc., 1955, 77, 2626.
8. D. Husain and R. G. W. Norrish, Proc. R. Soc. London, 1963,
    273A, 145.
9. T. Takeyama and H. Miyama, J. Chem. Phys., 1965, 42, 3737.
10. D. C. Bull, Combustion and Flame , 1968, 12, 603.
11. J. N. Bradley, R. N. Butlin, and D. Lewis, Trans. Faraday
    Soc., 1968, 64, 71.
12. L. J. Drummond, Combustion Science and Technology, 1972,
    5, 175.
13. R. K. Jayanty, R. Simonaitis, and J. Heicklen, J. Phys. Chem.,
    1976, 80, 433.
14. W. Hack, O. Horie, and H. Gg. Wagner, J. Phys. Chem., 1982,
    86, 765.
15. P. B. Pagsberg, J. Eriksen, and H. C. Christensen, J. Phys.
    Chem., 1979, 83, 582.
16. S. W. Benson, 18th Symp. (Intl.) Combustion Proc., 18th,
    1981, 882.
17. C. Pouchan and M. Chaillet, Chem. Phys. Lett., 1982, 90,
    310.
18. R. Lesclaux and M. Demissy, Nouv. J. Chim., 1977, 1, 443.
19. S. G. Cheskis and O. M. Sarkisov, Chem. Phys. Lett., 1979,
    62, 72.
20. C. Zetzsch and I. Hansen, Ber. Bunsenges Phys. Chem., 1978,
    82, 830.
21. J. C. McConnell, J. Geophys. Res. 1973, 78, 7812.
22. M. Gehring, K. Hoyermann, H. Schacke, and J. Wolfrum, 14th
    Symp. (Int.) Combust. Proc. 14th, 1973, 99.

23. E. A. Albers, K. Hoyermann, H. Gg. Wagner, and J. Wolfrum, 12th Symp. (Int.) Combust. Proc., 12th, 1969, 313.

24. R. Krishnan, M. J. Frisch, and J. A. Pople, J. Chem. Phys. 1980, 72, 4244; R. Krishnan and J. A. Pople, Int. J. Quant. Chem., 1978, 14, 91.

25. J. S. Binkley and C. F. Melius, to be published.

26. P. E. Hariharan and J. A. Pople, Theor. Chim. Acta, 1973, 28, 213.

27. C. C. J. Roothaan, Rev. Mod. Phys., 1951, 23, 19.

28. J. A. Pople and R. K. Ne sbet, J. Chem. Phys., 1954, 22, 571.

29. J. A.Pople, R. Krishnan, H. B. Schlegel, and J. S. Binkley, Int. J. Quant. Chem., 1979, 513, 225.

30. P. J. Hay, T. H. Dunning, and W. A. Goddard, J. Chem. Phys., 1975, 62, 3912.

31. K. Yamaguchi, S. Yabushita, and T. Fueno, J. Chem. Phys., 1979, 71, 2321.

32. A. J. Varandas and J. N. Murrell, Chem. Phys. Lett., 1982, 88, 1.

33. L. B. Harding and W. A. Goddard, J. Amer. Chem. Soc., 1978, 100, 7180.

34. P. J. Bruna, Chem. Phys., 1980, 49, 39; P. J. Bruna and C. M. Marian, Chem. Phys. Lett, 1979, 67, 109.

35. O. Nomura and S. Iwata, Chem. Phys. Lett., 1979, 66, 523.

RECEIVED December 21, 1983

DETONATIONS AND IGNITION

Fast Flames and Detonations

JOHN H. S. LEE

McGill University, Mechanical Engineering Department, 817 Sherbrooke St. W., Montreal,
Quebec, H3A 2K6, Canada

This paper centers on the problem of turbulent flame
acceleration by obstacles and the prediction of the
dynamic detonation parameters of fuel-air mixtures.
Current state-of-the-art in the understanding of
these phenomena, as well as progress made in achieving empirical and quantitative descriptions of these
combustion processes, are reviewed. The specific
topics discussed are i) the maximum attainable turbulent flame speed in an obstacle array, ii) computer simulation of turbulent flame accelerations,
iii) correlation between the detonation cell size
and the dynamic parameters of fuel-air detonations,
and iv) the transition from deflagration to detonation. Future directions in the investigation of
these problems are also discussed.

The primary concern of engineers involved in the design of nuclear
reactors, chemical plants, off-shore platforms and tankers, is the
overpressure-time development when large volumes of fuel-air mixtures are accidentally formed and ignited. The problem of estimating this overpressure-time development is an extremely complex
one. Depending on the initial and boundary conditions, the same
combustible mixture can yield volumetric or mass burning rate
that differs by a few orders of magnitudes. For example, the
burning velocity of laminar flames in stoichiometric mixtures of
the common hydrocarbon fuels with air is of the order of half a
meter per second, corresponding to flame speeds (i.e., relative to
the fixed observer) of about a few meters per second. However,
under suitable conditions, the same mixtures can sustain fast turbulent flames with flame speeds of the order of hundreds of meters
per second. Quite often, these fast flames transit spontaneously
to detonations which propagate to a typical velocity of about 1800
m/sec in these stoichiometric fuel-air mixtures. The overpressure-time development depends on the mass burning rate or the
flame speed which, in turn, depends strongly on the boundary
conditions (mainly the confinement).

0097-6156/84/0249-0119$09.00/0
© 1984 American Chemical Society

In closed vessels (fully confined), the eventual peak pressure obtained depends primarily on the energetics of the mixtures. The burning rate controls to some extent the amount of heat (hence the pressure loss) during the combustion process itself and this will result in a lower value for the final peak overpressure. The peak overpressure for confined explosions corresponds closely to the theoretical constant volume or isochoric explosion pressure for the particular mixture, which can be calculated readily from equilibrium thermodynamics when losses are ignored. The estimation of the rate of overpressure rise, however, is much more difficult since it depends on the mass burning rate or flame speed. For slow laminar flames, the rate of pressure rise is slow and venting is often quite effective in limiting the peak overpressure to any desired level. In unconfined conditions, the overpressure rise associated with slow deflagrations are usually negligibly small.

Under the appropriate boundary conditions and confinement, slow deflagrations can accelerate rapidly to very fast turbulent deflagrations. Turbulent flames depend strongly on the boundary conditions which determine the magnitude of the displacement flow velocity and hence the turbulence level in the unburned mixture. Thus, turbulent flame speeds may range from slightly above laminar flame speeds of a few meters per second to a few hundreds of meters per second just prior to the onset of detonation. The rate of overpressure rise is strongly dependent on the flame speed. For fast deflagration above a few tens of meters per second, venting becomes ineffective. Confinement also plays a decreasingly minor role in controlling the peak overpressure achieved. For violent turbulent flames with supersonic speeds of the order of a few hundreds of meters per second, overpressures of the order of a few bars are obtained even in unconfined geometry.

Fast turbulent deflagrations often transit spontaneously to detonations. For fully developed self-sustained detonation, boundary conditions and confinement play minor roles. The Chapman-Jouguet velocity and overpressure are based on the energetics of the mixture and can be evaluated from equilibrium thermodynamic computations. During the onset of detonation, the transient peak overpressures developed can be much higher than the equilibrium detonation pressures. Transition from deflagration to detonation is to be avoided whenever possible because of this extremely high pressure transient at the onset of detonation.

The prediction of the overpressure-time history associated with the combustion of an explosive mixture under specified conditions is the central problem of research in the loss prevention field. In the past decade, significant progress has been made in the understanding of the fundamental mechanisms involved in the complex combustion processes. Prompted by the urgency in resolving safety issues in LNG transport, offshore oil production platforms, and nuclear reactors, extensive research programs on gas

explosions involving large scale field experiments have been carried out in various countries. A number of reviews have already
been written describing the important progress made in this field
(1-4). The present paper emphasizes the more recent results, particularly the author's own work, in this very rapidly advancing
field. However, a certain amount of fundamental ideas exposed in
previous articles will be repeated to make the paper self-con-
tained. The earlier review papers, particularly References 1 and
2, should be referred to for a complete discussion of the turbulent flame acceleration mechanisms.

Turbulent Flame Acceleration (General Considerations)

It has been established that by far the most important mechanism
of flame acceleration is that due to turbulence created by obstacles in the path of the propagating flame. As a result of the
large temperature change across a flame, there corresponds an increase in the specific volume. This leads to a displacement flow
in the unburned mixture ahead of the propagating flame with a velocity of about the same order of magnitude as the speed of the
propagating flame itself. Large physical obstructions in this
displacement flow will create a velocity gradient field as well as
shear layers. As the flame advances into this velocity gradient
field, the flame sheet will be "stretched" and "folded" resulting
in an enlargement of the burning area and thus increasing the
volumetric or mass burning rate. Turbulence in the shear layers
in the wake of the obstructions also enhances the local burning
velocity of the "folded" flame sheet because of the higher turbulent transport rates. This combined effect of large scale flame
folding and finer scale turbulent enhancement of the local burning
velocity contributes to increase the volumetric burning rate.
Hence, a higher effective "global" turbulent flame speed is obtained. Higher flame speeds will in turn lead to higher displacement flow velocities in the unburned mixture, which then give rise
to more severe velocity gradients and higher turbulent intensity
in the shear layers. This positive feedback loop constitutes a
very powerful flame acceleration mechanism. Flame speeds of the
order of hundreds of meters per second and even spontaneous transitions from deflagration to detonation have been observed in
rather insensitive fuel-air mixtures under appropriate conditions.
Pioneering studies of flame acceleration in an obstacle array were
carried out by Chapman and Wheeler (6) using orifice plates in a
round tube and by Shchelkhin (7), who used a coiled wire helix inside the tube to create the turbulence. In the past decade, extensive experiments on flame acceleration have been carried out at
Göttingen and McGill University. The investigations at the Insti-
tut für Physikalische Chemie in Göttingen have been reviewed by
Wagner (8), while the studies carried out at the Shock Wave
Physics Laboratory at McGill have been summarized by Moen (4) in
1981. Large scale experiments in a 2.5 m diameter by 10 m long

tube have also been carried out in Norway and these studies have
been reviewed by Hjertager (9). Fairly large scale experiments on
flame accelerations have also been reported by Geiger (10) of the
Battelle Institute in Frankfurt and by Zeeuwen (11) of the TNO in
Holland. Urtiew (12) and Strehlow (13) have also been carrying
out some laboratory scale studies on obstacle induced flame accelerations and a major large scale facility called FLAME (a rectangular reinforced concrete channel 1.8 m x 2.4 m x 30 m), has
recently been completed at Sandia National Laboratory in New
Mexico for flame acceleration studies in H_2-air mixtures (14) in
connection with light water nuclear reactor safety research.

It may be concluded that all these studies on turbulent flame
acceleration by repeated or periodic obstacles have thus far produced essentially the following results. They demonstrate that
flames, even in rather insensitive fuel-air mixtures, can, under
appropriate conditions, accelerate very rapidly to very high
speeds and develop very high overpressures. The acceleration rate,
the flame speed attained and the associated overpressures developed are all functions of the particular fuel, the mixture composition, details of the obstacle configurations and the confinement
and geometry of flame itself. On a qualitative basis, the flame
acceleration mechanisms are fairly well understood. On a quantitative basis, "a-priori" predictions of the turbulent flame speed
and pressure development with given initial and boundary conditions are still in the far distant future. The two formidable
problems that have to be overcome are i) the predictions of the
transient transmissible turbulent flow structure in the unburned
mixture ahead of the flame, and ii) the predictions of the combustion processes for any given turbulent flow structure. Computer
modelling with appropriate empirical input from experiments may
play an important role in providing an analytical tool for quantitative predictions within a limited range of experimental conditions. The kind of useful experimental information would be maximum steady state values for the turbulent flame speeds in given
obstacle arrays. The maximum turbulent flame speed represents a
critical balance between the positive (enhanced turbulent diffusi-
vities) and negative (quenching) mechanisms of turbulent combustion. Information on the dependence of this maximum flame speed
on the type of fuel and mixture composition (hence the chemical
kinetics) and the obstacle configurations (turbulent flow structure) may be used to construct the kind of empirical relationships
needed for these computer models. Most of the experimental studies thus far, particularly the large scale experiments, have been
carried out in rather "short" apparatus, where the length of the
flame travel is not large, as compared to its transverse dimension
(i.e., L/D ratio for tube). Thus, only the initial transient
phase or the acceleration process is observed. In the past few
years, an extensive experimental program was initiated to investigate the dependence of the maximum steady state turbulent flame
speed on fuels, composition and obstacle configurations. This
paper presents some of the results obtained to date.

Turbulent Flame Acceleration (Recent Results)
==

To obtain the maximum steady state turbulent flame speed requires
relatively long flame travel. Three steel combustion tubes of
about 12 m long and respective diameters of 5 cm, 15 cm, and 30 cm
are used. Even for the largest 30 cm tube, the flame propagation
can be observed over a distance of about 40 tube diameters. The
tubes are closed at both ends, with ignition effected by a spark
at one end. For about 80% of the flame travel, the boundary conditions at the other end of the tube do not play any significant
roles. The flame propagation is essentially the same as that for
a tube with an open downstream end. Premixed mixtures of various
fuels (C_2H_2, H_2, C_2H_4, CH_4, C_3H_8) with air are filled into the tube
through standard procedures. Diagnostics consist of time of ar-
tival, ionization gauges (spaced about 0.5 m apart) for flame
speed measurements and piezoelectric transducers (at varying locations along the tube) for monitoring the pressure development. It
was decided to use one standard type of simple obstacle configuration (i.e., circular orifice plates of various blockage ratios,
$BR = (1 - [d^2/D^2])$, so that the flow structure is relatively
straightforward and symmetrical, thus permitting analysis to be
carried out. In the 5 cm diameter tube, a "Shchelkhin spiral"
made out of a 6.3 mm diameter copper tubing with a pitch of 5 cm
is also used for comparison purposes.

In general, the flame upon ignition will accelerate rapidly
and reach steady state within a few meters of flame travel for
most of the obstacle configurations studied. Little attention
thus far has been devoted to the analysis of the rate of acceleration. The steady state flame speed is determined by averaging
over the speeds obtained from a number of different ionization
probes. Under certain conditions, the steady state flame speed
may show large fluctuations. To demonstrate the general features,
the results for H_2-air mixtures in the 5 cm tube using a 3 m
length of Shchelkhin spiral (BR = 0.44) is shown in Figure 1. It
can be observed that the flame accelerates to a maximum steady
state value in about 1 m for a wide range of H_2 concentration.
This maximum steady state turbulent flame speed depends also on
the H_2 concentration itself and ranges from about 100 m/sec ($\sim$ 10%
H_2) to about 1800 m/sec (30% $\lesssim$ H_2 $\lesssim$ 45% H_2). Upon exiting the 3 m
length of the obstacle (i.e., the Shchelkhin spiral) into the
"smooth" tube, the flame immediately decelerates. For H_2 $\lesssim$ 15%,
the flame decelerates to a new and much lower steady state value
corresponding to the smooth wall boundary condition. However, for
H_2 $\gtrsim$ 15%, the flame reaccelerates and transits to detonation after
a couple of meters of flame travel (the transition distance depends on the H_2 concentration). We shall elaborate more on the
transition phenomenon in a later section.

The maximum steady-state turbulent flame speeds as a function

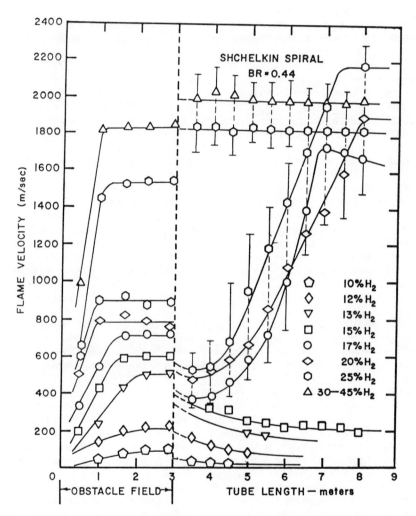

Figure 1. Variation of flame speed with distance for H_2-air mixtures in a 5-cm diameter tube with a 3-m length of Shchelkhin spiral obstacle.

of H_2 concentration for the 5 cm and 15 cm diameter tubes using
circular orifice obstacles of BR = 0.44 and 0.39 spaced one tube
diameter apart are shown in Figure 2. The results indicate that
for the case of H_2-air mixtures, the dependence on tube diameter
as well as blockage ratios is not very strong. This suggests that
for this very sensitive fuel, the 5 cm scale is already sufficiently large when compared to the chemical length scale of the
mixture for the combustion processes to assert a significant scale
dependence on the tube diameter. Similar remarks apply to the
dependence of the flame speed on the blockage ratio. For the
"roughness" scale associated with BR = 0.44 and 0.39 (or D-d $\simeq$ 1.6
cm and 0.63 cm for the 15 cm and 5 cm tube respectively), the chemical length scale for the sensitive H_2-air mixture is also sufficiently small for the flame to have a strong dependence on the
range of the blockage ratio of 0.39 $\lesssim$ BR $\lesssim$ 0.44. However, stronger dependence on scale would be anticipated for the very lean H_2-
air mixtures where the chemical length scale would be larger. The
dependence of the flame speed on mixture composition is, however,
much more pronounced. For H_2 $\lesssim$ 13%, flame speeds in both tubes
are of the order of about 100 m/sec. Then a very sharp "jump"
occurs at H_2 $\simeq$ 13% when the flame speed increases abruptly by
three or four fold. This abrupt increase may be credited to the
change in the chemical length scale. As pointed out by Atkinson
(15) among others, the reaction H + O_2 $\rightarrow$ OH + O is in competition
with the three body reaction, H + O_2 + M $\rightarrow$ HO_2 + M with the latter
being more dominant for temperatures below 1300°K. At about 13%
H_2, the adiabatic flame temperature is of the order of 1300°K.
Thus, a changeover to the more rapid OH branching reaction may
take place at H_2 $\simeq$ 13% and give rise to a more rapid increase in
the flame speed. The chemical length scale is also significantly
reduced when the kinetic mechanism changes to rapid OH branching.
    A second rapid increase in the flame speed occurs at H_2 $\gtrsim$ 25%
for both tubes. This corresponds to a transition to the detonation regime. The detonation velocity in the obstacle field is
typically about 1500 m/sec and is practically independent of the
H_2 concentration up to H_2 $\simeq$ 45%. For higher H_2 concentrations,
the detonation velocity abruptly drops back to the values for deflagration speeds of the order of 800 m/sec. The severe pressure
(or momentum) losses due to the presence of the obstacles accounts
for the sub-Chapman-Jouguet detonation velocities observed. The
normal velocities are about 2000 m/sec as observed in smooth tubes
for H_2 concentrations in the range (i.e., 25% $\lesssim$ H_2 $\lesssim$ 45%).
    In the range of 13% $\lesssim$ H_2 $\lesssim$ 25%, the flame speed showed a con-
tinuous steady increase with the H_2 concentration. This regime
is probably dominated by the pressure losses and gasdynamic "chok-
ing" at the orifice. A higher rate of energy release (i.e., flame
power) accounts for the steady continuous increase of the turbulent flame speed with fuel (i.e., H_2) concentration. Preliminary
experiments in the 30 cm diameter tube obtained recently showed
similar qualitative results.

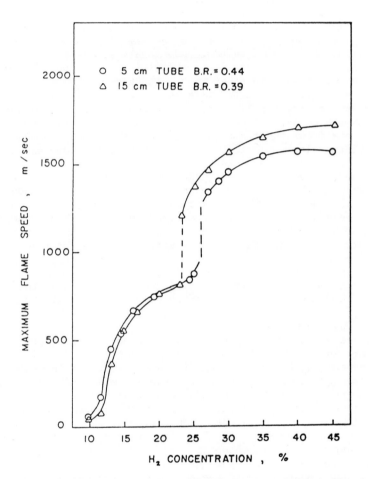

Figure 2. Comparison between the maximum turbulent flame speed
for H_2-air mixtures in the 5-cm and 15-cm diameter tube with
orifice plate obstacles.

The peak overpressure associated with these steady state fast
deflagrations in H_2-air mixtures are shown in Figure 3. Similar
to the flame speed, the dependence on tube diameter and blockage
ratios are also minimal. The same qualitative trend is observed
for the dependence of the peak overpressure on the mixture composition. The magnitude of the peak overpressure is also compatible
with the theoretical estimate based on the flame speeds given in
Figure 2. Thus, these fast flames do correspond to very high mass
or volumetric burning rates and are not associated with the kine-
matics of the motion of the unburned mixtures ahead of the flame.
In fact, for the supersonic flames (i.e., flame speeds greater
than the sound speed in the unburned mixtures), the pressure-time
profiles from the piezoelectric transducers located at various
positions along the tube indicatea series of "pressure spikes"
corresponding to the arrival of the rather extended turbulent
burning zone. The structure of these fast turbulent flames is
similar to that observed by Oppenheim et al. (16) for the turbulent deflagrations just prior to the onset of detonation. To
visualize better these fast deflagrations, the readers should con-
sult the works of Oppenheim and co-workers who took extensive high
speed stroboscopic laser Schlieren photographs of these fast turbulent flames with exceptional clarity and time resolution.

For the other hydrocarbon fuels, the more sensitive ones
(i.e., C_2H_2 and C_2H_4) as well as C_3H_8 and CH_4 are studied in the
smaller 5 cm tube. The steady state flame speeds corresponding to
orifice plate obstacles of BR = 0.44 for C_2H_2, C_2H_4 and C_3H_8 are
shown in Figure 4 as functions of the mixture composition (i.e.,
equivalence ratio ϕ where stoichiometric composition corresponds
to ϕ = 1). For acetylene which is even more sensitive to fuel
than hydrogen, it is observed that ignition by the weak spark used
became erratic when the C_2H_2 concentration is less than about 3.5%.
However, when ignition is successful, the flame accelerates very
rapidly to steady state values of about 800 m/sec. Similar to H_2-
air mixtures, abrupt transition to the detonation regime occurs at
around 5.5% C_2H_2 (ϕ = 0.66). After that, the detonation velocity
shows a slight continuous increase with the C_2H_2 concentration.
For the less sensitive fuel ethylene (C_2H_4), quenching is observed
for C_2H_4 concentration less than about 4.5% ($\phi \simeq 0.67$). By quenching, it means that ignition is successful but following a rapid
acceleration, the flame is then suddenly quenched. The quenching
phenomenon involves the flame first accelerating to a sufficiently
high velocity after the passage through a number of orifice plates,
then as the flame speed reaches some critical value, the turbulent
mixing rate becomes sufficiently fast to cool the burning zone as
the flame emerges from the orifice. Extinction then results. The
quenching limit depends on the orifice diameter for a given fuel
and mixture composition. The study of this quenching phenomenon
has already been reported by Thibault et al. (17) and the quenching criterion is found to be similar to that for the flame stabilization by bluff bodies where the "blow-off" limit requires a

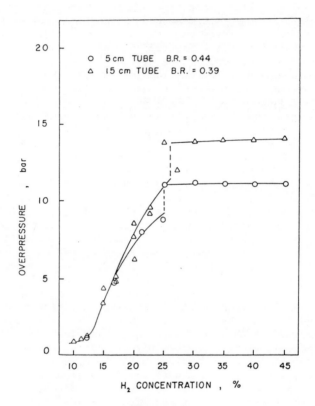

Figure 3.   Peak overpressures for turbulent H_2-air flames propagating in an orifice obstacle array.

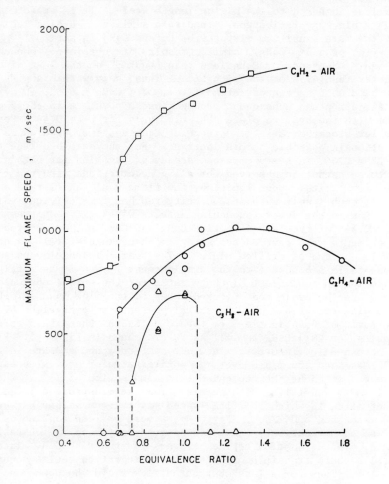

Figure 4.   Maximum turbulent flame speeds for C_2H_2-air, C_2H_4-air, and C_3H_8-air mixtures in a 5-cm diameter tube with orifice obstacles.

critical balance between the chemical length of the scale reacting
mixture and the fluid dynamic mixing length scale. For H_2 and
C_2H_2 where the quenching regime for the size of the orifice plates
used was not observed, the chemical length scale is sufficiently
small as compared to the mixing length scale. It is expected that
higher blockage ratios would lead to a similar quenching regime
for the more sensitive mixtures of H_2 and C_2H_2 in air.

For propane (C_3H_8), similar results are obtained. Quenching
is again observed for C_3H_8 less than about 3% on the lean side and
greater than 4.5% on the rich side. Thus, only within a very nar-
row band of C_3H_8 concentration (i.e., 3% $\lesssim$ C_3H_8 $\lesssim$ 4.5%) when flame
acceleration and subsequent propagation is possible in the 5 cm
tube with the orifice plate obstacles used for BR = 0.44. For a
smaller blockage ratio, the range of C_3H_8 concentration where propagation is possible should increase. For the same blockage ratio
of BR = 0.44, in the 5 cm tube, steady propagation for methane-air
mixtures cannot be observed for all concentrations with the flammability limits. Upon ignition, the flame would accelerate and
then quench itself when its velocity reaches some critical value.

Due to the severe quenching effects with orifice plates of
BR = 0.44 for the less sensitive C_3H_8 and CH_4 in the 5 cm tube,
experiments were carried out using a "Shchelkhin spiral" as obstacles instead of orifice plates. The spiral is continuous and
presents less blockage to the displacement flow. It is difficult
to define an equivalent blockage ratio for the spiral obstacle and
therefore the projected "blocked" area to the flow is used. For a
wire diameter of 6.3 mm, the spiral presents an equivalent blockage ratio of 0.44 in the 5 cm tube. Using the Shchelkhin spiral,
results for CH_4, C_3H_8 as well as H_2 are shown in Figure 5. Flames
in methane-air mixtures are now not self-quenched and are in general less than 400 m/sec over the entire range of equivalence ratio.
However, at about the stoichiometric composition of $\phi \simeq 1$, where
it is most sensitive, methane-air flames demonstrate a large
fluctuation in which the flame speed may take on values between
400 m/sec and 800 m/sec. For C_3H_8, a more or less continuous increase to detonation speeds are observed in the spiral obstacle
field. The detonation regime persists for the rich C_3H_8 mixtures
until a sharp transition back to the deflagration regime occurs
$\phi \gtrsim 1.6$. Thus for a less obstructed flow field, both CH_4 and C_3H_8
demonstrate the existence of high steady state turbulent flame
speeds, although the magnitude of the flame speed is slightly less
than the corresponding values for the more sensitive fuels. The
importance of the flame speed scale is apparent for less sensitive
mixtures with larger chemical length scales. For H_2-air mixtures,
results within the spiral obstacle are similar to that for the
orifice plate obstacles. With less blockage, hence pressure
losses, the detonation velocities are now close to the theoretical
Chapman-Jouguet values.

For the less sensitive mixtures, the scaling up to large
dimensions (i.e., tube diameter) becomes important. Figure 6

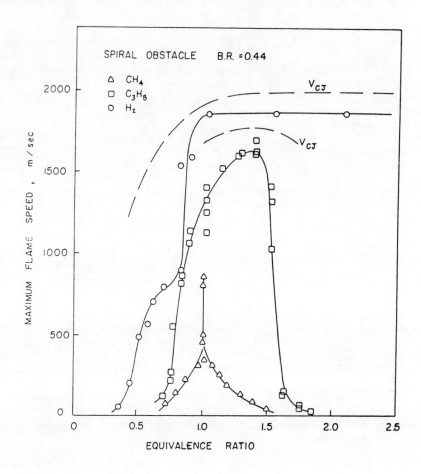

Figure 5. Comparison of the maximum flame speed for CH_4, C_3H_8, and H_2 in a 5-cm tube with a Shchelkhin spiral obstacle.

CHEMISTRY OF COMBUSTION PROCESSES

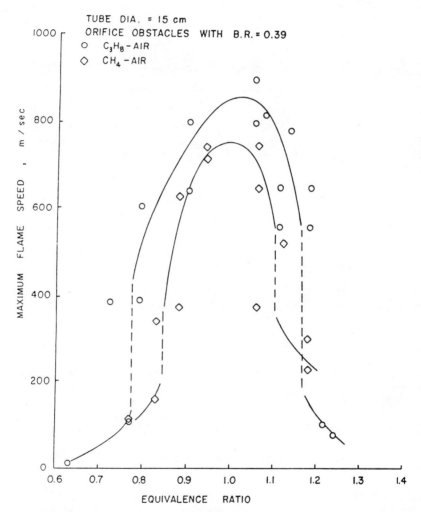

Figure 6. Maximum turbulent flame speeds for C_3H_8-air and C_4-air in the 15-cm diameter tube with orifice obstacles.

shows the results for CH_4 and C_3H_8 in the 15 cm tube with orifice
plate obstacles of BR = 0.39.   For C_3H_8, sudden transition to high
speed supersonic deflagrations occurs at equivalence ratios $\phi \simeq$
0.75 and $\phi \simeq 1.2$.   For $\phi \lesssim 0.75$ and $\phi \gtrsim 1.2$, typical flame speeds
are of the order of 100 m/sec and below.   Maximum flame speed oc-
curs at around stoichiometric composition of $\phi \simeq 1$.   However, even
for the case of propane, transitions to detonations are not observed.   Similar results are obtained for methane, but the range of
composition in which steady state propagation without quenching is
slightly narrower than propane.   For example, the transition to
the supersonic regime now occurs at $0.85 \lesssim \phi \lesssim 1.1$ instead of
$0.75 \lesssim \phi \lesssim 1.2$ for propane.   Maximum flame speeds for CH_4 at $\phi \simeq 1$
is about 750 m/sec, which is slight lower than the corresponding
value for propane of about 850 m/sec.   It should be noted that
these maximum flame speeds are compatible to those reported by
Moen et al. (18) and Hjertager (19) in a much larger scale apparatus (2.5 m diameter and 10 m long).   However, with an L/D ratio of
4 it is doubtful that the flame speed values reported by Moen and
Hjertager do represent the final steady state values.   Current experiments in the 30 cm tube will elucidate further on the important scaling problem.

Turbulent Flame Acceleration (Computer Simulation)

There exist numerous computer codes capable of simulating multi-
dimensional time dependent turbulent reacting flows.   A number of
attempts have already been made to adapt these computer codes to
model the problem of turbulent flame acceleration by repeated obstacles (20-22).   With the appropriate input from experiments,
they all managed to reproduce reasonably well the trend of the experimental results.   Due to the expenses involved in carrying out
large full scale experiments, computer modelling or simulation
will play an important role in providing estimates of pressure de-
velopments in full scale situations under a variety of initial and
boundary conditions, based on input from a few selected scaled
down experiments.   These computer codes are in general, sufficiently complex so that they are not accessible to those outside
the group that developed them.   At any rate, acquiring the exper-
tise required to use any one of these codes would take a commit-
ment of one or two man-years.   Depending on the accuracy desired,
the cost in running these computer codes is not trivial and is
quite often comparable to the cost of the actual performance of
scaled down experiments.   Thus, except for Government Institutions
and large Research Centers, it is not likely to see a widespread
use of these complex computer codes.

   It is of interest to review some of the results produced by
one of these computer codes.   Due to the high running cost, the
Vortex Dynamics Code (22) would not likely be an attractive candi-
date to use in simulating actual complex situations.   The "k-ε"
type code of Hjertager (23) or the CONCHAS-SPRAY code recently

adapted to the flame acceleration problem by Marx (24) would be more appropriate for such purposes. However, the Vortex Dynamics Code gives an accurate description of the turbulent flow structure and the behaviour of the flame surface as it propagates through the vorticity field. Thus, the Vortex Dynamics Code can be used to supplement the fundamental studies on flame accelerations by providing a realistic picture of the development of the flame structure in the complex turbulent field. In other words, the Vortex Dynamics Code provides a high speed schlieren movie of the complex turbulent flame structure to facilitate the interpretation of the actual pressure-time and flame speed data from experiments. The actual taking of high speed schlieren movies in flame accelera- tions in atmospheric fuel-air mixtures is not feasible due to the high pressure and large scales involved. Then the Vortex Dynamics Codes can be looked at as more of an experimental diagnostic tech- nique than an algorithm for estimating the pressure development of flame accelerations in complex situations.

Figure 7 shows a selected sequence of a computer generated movie using the Vortex Dynamics Code of a flame propagating in a two dimensional channel with baffle plates as obstacles spaced one channel width apart. As the flame advances from the closed end (ignition) of the tube, the displacement flow through the orifice creates shear layers as indicated by the regions of concentrated vorticities. On the outflow through the orifice, a velocity grad- ient is generated upstream of an orifice and the distortion of the flame as it propagates into this gradient field is clearly appar- ent in the fourth frame of Figure 7. The very rapid stretching of the flame as the flame "tip" is being convected by the jets through the orifice is illustrated in the last frame. It should be noted that in this sequence of computations, the flame sheet is treated as a thin surface with constant local burning velocity. When a dependence of the local burning velocity on the turbulence field is induced in the computations, a much more rapid burning will occur in the shear layers. The computer generated movie shown in Figure 7 corresponds very closely to the actual high speed schlieren photographs of flame acceleration through a baffle obstacle (25).

The Vortex Dynamics Code has been applied recently to the study of flame acceleration in a multiple compartment chamber (26). Figure 8 shows a sequence from the computer generated movie using the Vortex Dynamics Code. Again, the generation of vorticity in the shear layers in the wake of the obstacles is clearly illus- trated. The stretching of the flame until burning occurs simul- taneously in all the compartments can also be observed. Of parti- cular interest to note is that vorticity generated in upstream shear regions is continuously convected downstream by the flow. Thus, as the flame advances, it will be propagating into an in- creasingly more intense vorticity field, which is due to both the generation that occurs locally at the shear layers, plus that con- vected from upstream shear regions. This leads to very intense

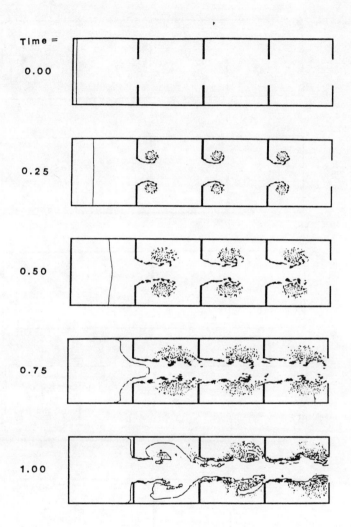

Figure 7. Computer simulation of flame   acceleration in an array
of orifice obstacles using the Vortex Dynamics Code.

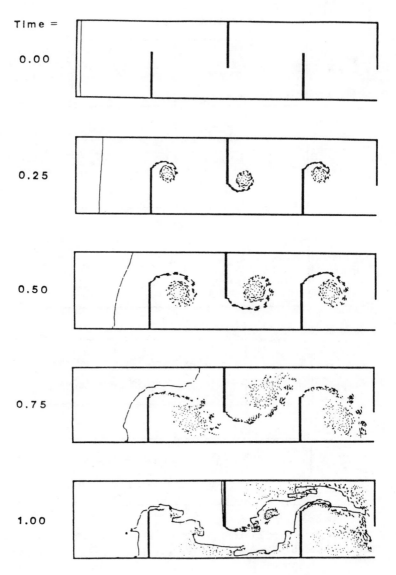

Figure 8. Computer simulation using the Vortex Dynamics Code for flame acceleration in multiple compartment chambers.

turbulence and thus very high burning rates. This mechanism of
local burning rate enhancement due to vorticity swept down from
upstream shear regions is very important. It may result in a decreasing dependence of the flame acceleration rate on the local
boundary conditions. This also implies that very high burning
rates can occur in the gases that are vented from one region to
another due to the vorticity contained in them. The above discussions illustrate the kind of combustion to the fundamental
understanding of flame acceleration that computer simulation can
provide. To make actual turbulence measurements experimentally
for these accelerating flames and perform the necessary data reduction and analysis to achieve the same kind of quantitative description of the turbulent field provided by the Vortex Dynamics
Code would be a formidable task. Thus, even though the simulation
thus far has neglected important compressibility effects and burning rate-turbulence coupling, much can be learned already from the
fluid dynamic processes that it reproduces.

Detonations

Although not a common occurence for fuel-air mixtures, particularly under open unconfined conditions, detonations cannot be ignored in risk analysis. The distinctive capabilities of detonation waves necessitate the need to acquire a complete understanding of the phenomena so that the precise conditions in which a detonation wave can be initiated and propagated are known. In this
way, a realistic assessment of the detonation hazards associated
with a given installation can be obtained.
     In general, detonation waves in stoichiometric mixtures of
most of the common hydrocarbon fuels with air propagate at about
1800 m/sec with an overpressure increase of about 15 bars across
the wave. It is a curious fact of nature that the same explosive
mixture that normally burns with a laminar flame speed of about a
few meters per second can also support the intense combustion associated with these supersonic detonation waves. The processes
that go on inside the structure of detonation fronts are extremely
complex, involving multi-dimensional shock interactions in an intense turbulent reacting medium. However, it is also a curious
fact of nature that on the basis of the calculations using the one-
dimensional Chapman-Jouguet model with equilibrium thermodynamics
which ignore the complex structure, the predicted detonation properties (i.e., detonation velocity, overpressure, etc.) are found
to be quite close to that observed experimentally, even near the
detonation limits. On the other hand, the dynamic detonation
parameters (eg., detonability limits, initiation energy, critical
tube diameter, etc.), cannot be obtained from this equilibrium
Chapman-Jouguet theory. The complex dynamic processes that go on
inside the structure of the detonation wave (i.e., the propagation
mechanism) must now be taken into consideration if these dynamic
detonation parameters are to be predicted. Thus far, no theory

exists that can give even semi-quantitative estimates of these
dynamic parameters. Computer simulations using multi-dimensional
reactive shock codes have been demonstrated by Fujiwara (27) and
Oran (28). Both groups have reproduced two-dimensional cellular
detonations from their respective computer codes. However, the
computation time required is extremely long for the required time
and spatial resolutions. It is not likely that extensive computer
simulations of the detonation phenomena will be made. It is also
not clear at present just what useful information one can derive
from these computer simulations to justify the high cost of running them for the detonation problem.

On the experimental side, significant advances have been made
in the early sixties in the understanding of the fundamental processes that go on inside the detonation front itself. However,
this basic knowledge acquired was not exploited until the late
seventies when the detonation cell size was recognized as a useful
fundamental parameter that represents the chemical length scale of
the detonation process, and hence can be correlated to all the
important dynamic parameters. The measurement of the cell size
using carbon soot deposited on a thin polished metal foil inserted
into the detonation tube has been well established in the early
sixties. However, cell size measurements in fuel-air mixtures
were not revived until the late seventies by Bull (29) and
Knystautas (30,31) when the usefulness of cell size data became
apparent. The first step in establishing the correlation between
cell size and the dynamic parameters was made by Edwards (32) who
suggested that the correlation between the cell size "λ" and the
critical tube diameter "d_c" (i.e., $d_c = 13\ \lambda$) first observed by
Mitrofanov and Soloukhin (33) in 1964 should be universal. The
generality of the simple law $d_c = 13\ \lambda$ was confirmed by the extensive experimental work of Knystautas (30,31) who measured simultaneously the cell size as well as the central tube diameter for a
number of fuels over a range of fuel-air concentrations and initial pressures.

The $d_c = 13\ \lambda$ law led the author to develop a simple model
whereby the critical energy for initiation can be predicted when
the cell size and the equilibrium Chapman-Jouguet detonation
states are known. This simple so-called surface energy model of
Lee (34) gave predictions for the critical initiation charge
weight for various hydrocarbon fuel-air mixtures in close accord
with the experimental data obtained by Elsworth (39).

The relationship between detonability limits and the cell
size is based on a suggestion made by Wagner (35) in 1960, who
first stated that the appearance of the single headed spin wave
structure should correspond to the detonability limit in the given
tube. Since then, this suggestion of Wagner was demonstrated to
be true by Donato (36), who showed experimentally that for mixture
compositions outside the limit when the single head spin structure
first appeared, a detonation when disturbed will fail and will not
re-transit back to the detonation mode. Recent work by

Vasiliev (37) also confirms this criterion for detonation limit
based on the onset of the single headed spin structure. Thus,
with the cell size as a function of mixture compositions known,
the detonability limits in tubes can readily be established. The
criterion is that the limiting mixture composition should give a
cell size λ which is of the order of the tube diameter itself.
Limit criteria for other tube geometries and for unconfined conditions have yet to be established. However, there is no doubt
that the limit conditions should be correlated to the fundamental
chemical length scale of the mixture (i.e., cell size).

   A thorough review of the dynamic parameters of gaseous detonation waves and their correlations with the cell size λ has been
given by the author recently (38). However, for the sake of com-
pleteness, some of the more important results mentioned in the
above discussions will be given. Figure 9 shows the detonation
cell size for the common hydrocarbon fuels as a function of equivalence ratio (ϕ = 1 corresponds to stoichiometric composition).
They all take on the shape of U-curves with the minimum cell size
occurring at about the stoichiometric composition. Since the sensitivity of the mixture is inversely proportional to the cell size,
we see from Figure 9 that acetylene is the most sensitive fuel
with a minimum cell size λ = 0.5 cm, while methane, which has a
minimum cell size about two orders of magnitude larger (λ = 33 cm)
is the least sensitive of the common fuels. The rest of the
alkane family (C_3H_8, C_2H_6, C_4H_{10}) all have about the same cell
size and hence similar sensitivity. The order of sensitivity to
detonation for the various fuels as indicated by their respective
cell size in Figure 9 are in accord to that established by Matsui
and Lee (39) who devised a detonation hazard number based on the
critical initiation energy with the most sensitive fuel of C_2H_2 as
the reference. The solid curves in Figure 9 represent the correlation λ = Aℓ where "ℓ" is the induction zone length computed theoretically based on the detailed chemistry of the mixture (40).
The proportionately constant A is obtained by fitting at ϕ = 1.
The general U-shaped behaviour is reproduced but the error can be
quite significant for off stoichiometric mixtures. The correlation to detailed chemistry in a simple λ = Aℓ relationship is not
of sufficient accuracy for prediction purposes. In other words,
using one experimental data point to determine A and compute the
cell size henceforth using detailed chemistry is not accurate
enough for practical purposes. This is due to the fact that most
dynamic detonation parameters are sensitive functions of the cell
size (eg., $E_c \sim \lambda^3$).

   Based on the cell size data obtained experimentally (Figure
9), the critical tube diameter "d_c" can be obtained by simply mul-
tiplying λ by a factor of 13. Figure 10 shows a comparison between the d_c = 13 λ correlation and the experimental results from
the direct measurements of "d_c" itself from large scale experiments. The agreement is quite good in view of the difficulties in
controlling closely the experimental parameters in large scale
field experiments.

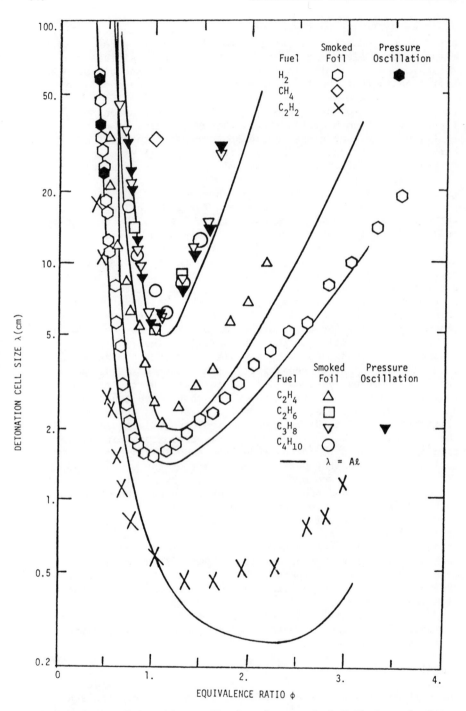

Figure 9.   Detonation cell size from smoked foil records for fuel-air mixture.

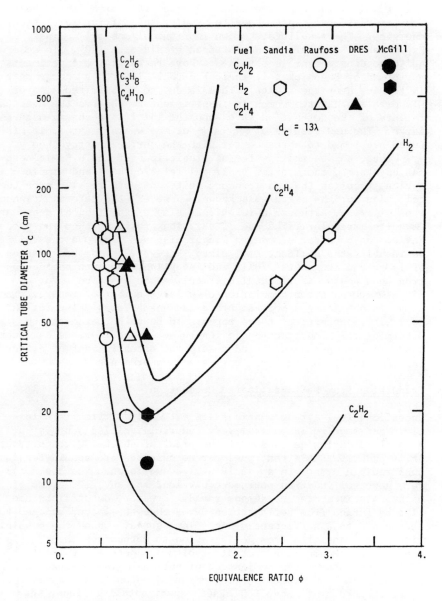

Figure 10. Comparison of the critical tube diameter with the $d_c = 13 \lambda$ correlation.

Based on the experimentally obtained cell size data, the surface energy model can be used to predict the initial initiation
energies. The results are shown in Figure 11 as solid lines.
Experimental data are those obtained by Elsworth (40). In general,
quite good agreement is obtained. More recent large experiments
performed by Benedick (41) indicated that Elsworth's energy data
is too low near the limits. This is due to the limited size of
Elsworth's apparatus which is insufficient to observe the eventual
failure of the detonation wave when the initiation charge used is
large. The new results of Benedick for the weaker near limit mixtures are found to agree better with the surface energy theory.

Thus, we see that as far as predictions of the dynamic detonation states, there exist empirical relationships and simple analytical theories that can produce quite adequate results once the
cell size for the mixture is known. However, the determination of
the cell size, although a relatively simple experimental procedure,
is quite irregular. The use of very long foils so that the propagation can be observed over a longer distance, improve the ability
to identify the dominant cell size. However, the smoked coating
of large and long metal foils and inserting them into the detonation tube poses experimental difficulties in practice. For a
given mixture, the accumulation of a large number of records also
helps in obtaining a more accurate and unambiguous value for the
cell size. At present there appears to be no better means to
determine this important basic length scale and there is no doubt
that efforts must be directed toward the establishment of better
experimental techniques for cell size measurements.

Transition from Deflagration to Detonation

Unconfined fuel-air detonations are extremely difficult to initiate and require in general very powerful ignition sources such
as a solid explosive charge of the order of about 100 grams. Thus,
it is understandable that fuel-air detonations are considered as
non-credible events in an accidental scenario when the absence of
very powerful ignition sources equivalent to, say, 100 grams of
high explosives, can be demonstrated. However, detonations can
also be formed from deflagration waves when the latter managed to
accelerate to a sufficiently high flame speed. Thus, the possibility of a transition from deflagration to detonation (D.D.T.
henceforth) becomes the central problem and determines if
detonation hazards can be ignored or not in a given situation. In
general, the flame speeds at the onset of detonation are found to
be of the order of about 800 m/sec experimentally. Thus, the
possibility of a DDT hinges on the question of flame accelerations.
If powerful flame acceleration mechanisms such as turbulence generated by repeated obstacles and the right kind of boundary conditions (confinement) are present, then DDT is a definite possibility. In long confined channels with a distribution of large

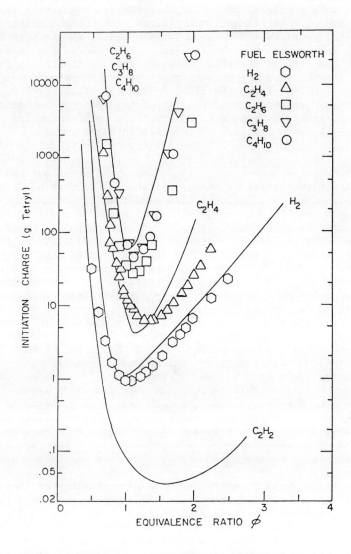

Figure 11. Comparison of theoretical predictions of critical initiation energy with Elsworth's experimental data.

obstacles in the flow, few would debate the possibility of a DDT. However, in a partially or totally unconfined environment or in the absence of a sufficiently long flame travel for the acceleration to the required flame speed to take place, DDT is then considered to be a highly improbable event. This conclusion is mainly based on the lack of a viable mechanism whereby the necessary conditions for the onset of detonation to occur can be produced without flame accelerations. In 1978, Knystautas et al. (42) reported that direct initiation of a spherical detonation can be achieved when a sufficiently intense turbulent jet of hot combustion products (of the same mixture) is injected into the unconfined cloud of explosive mixture. By direct initiation, it is meant that the detonation is formed "instantaneously" in the immediate vicinity of the ignition source (i.e., the turbulent jet) without any pre-detonation flame travel or run-up distance. In other words, it is no longer necessary to rely on flame acceleration mechanisms to bring about the DDT. As long as certain required conditions can be generated locally, which in this case is the rapid turbulent mixing between hot product gases and the unburned mixture, then onset of detonation results. This important experiment of Knystautas produced the much sought after mechanism for DDT to occur in an unconfined cloud. The possibility of a small local confined explosion which subsequently ruptured its containment and resulted in a venting of the hot product gases to the outside mixture is certainly a very common accident scenario in industrial environments. However, the experiments of Knystautas were carried out in sensitive oxygen enriched fuel-air mixtures and only for the case of the most sensitive fuel of stoichiometric C_2H_2-air when the turbulent jet induced DDT was he able to observe DDT. It is obvious that the scale of turbulent mixing region required for the less sensitive fuels would be much larger than the maximum size of a 20 cm diameter jet that was used by Knystautas. No large scale studies were followed up, and since then only fundamental studies were continued to learn more about the dependence of the chemistry on the mixing rate, turbulence scale and jet diameter.

There are three recent reports of large scale experiments where DDT was observed. In a flame propagation experiment in a partially confined rectangular channel, DDT was observed in H_2-air mixture (∿40% H_2) by Pfförtner (43). The local turbulent mixing region was caused by a small ventilation fan which was left on during the flame propagation. Thus, as the flame front arrives at the fan, the rapid mixing caused by the fan between hot gases in the flame zone and the unburned mixture is sufficient to result in the onset of detonation. Although 40% H_2 in air is a fairly sensitive mixture, Pfförtner's observation nevertheless confirmed the mechanism of mixing induced DDT in an unconfined cloud.

Recently, Geiger (44) of the Battelle Institute in Frankfurt also reported the observation of a DDT in unconfined stoichiometric H_2-air mixtures. In Geiger's experiment, a rectangular box

0.50 m x 0.50 m x 1 m, closed at the ignition end and connected
through a small orifice at the other end to a large volume of the
same explosive mixture contained in a 1 m x 1 m x 3 m plastic bag,
was used. The turbulent jet discharged from the orifice provides
a powerful ignition source. However, in this experiment DDT was
not observed in the jet mixing region. The powerful jet ignition,
however, led to the development of a very fast deflagration. DDT
was then later observed when the flame had propagated some distance down the bag. Again, the mixture used is also a fairly sensitive one.

Moen (45) also reported the observation of DDT in an unconfined volume of a much less sensitive off-stoichiometric mixture
of C_2H_2-air ($\sim 5\% \ C_2H_2$). The intention of Moen's experiment was
not to investigate the DDT phenomenon. However, a misfire of the
initiating charge resulted in flame propagation (instead of a
detonation) in the tube, which subsequently discharged into a
large volume of the same mixture contained in a large plastic bag
at the end of the tube. The tube is ~ 0.63 m diameter and 3 m long
and the plastic bag is 2 m diameter and 3.5 m long. From high
speed movie records, Moen was able to estimate the velocity of the
hot gases discharging from the tube, and found it to be about 550
m/sec. As in the case of Geiger's experiment, DDT again did not
occur at the immediate vicinity of the tube exit. The jet was observed to impinge on a vertical wall at the end of the plastic bag,
creating a recirculation zone (vortex) at the base (i.e., between
the wall and the ground). The onset of detonation was observed to
occur in this recirculation vortex and once formed, the detonation
propagated throughout the entire volume of mixture at about 1700
m/sec, corresponding quite closely to the anticipated Chapman-
Jouguet value.

There are three recent large scale experiments that reported
the occurence of DDT in unconfined mixtures. They all support the
turbulent jet mixing mechanism advanced by Knystautas et al. The
flame acceleration experiments reported earlier in Section 3 have
helped considerably to elucidate the necessary conditions required
for the onset of DDT. In the confined tube experiments it was
found that onset of detonation requires flame speeds of the order
of about 800 m/sec. This gives the minimum turbulent intensity
required for the mixing. The minimum size of the mixing region
(tube diameter) itself, was also found to be of the order of the
cell size. Translated to the unconfined environment, it appears
that DDT would occur if a sufficiently intense turbulent mixing
region of a size of the order of the dimension of the cell size
of the mixture could be formed. Once this initiation kernel is
produced, then onset of detonation results and the detonation will
subsequently propagate throughout the available unburned volume.

Much investigation, particularly large scale experiments, are
needed to clarify and define more precisely the required conditions postulate for DDT. However, it appears that this important
question can be answered satisfactorily in the near future.

Conclusions

Turbulent flame acceleration by an obstacle array should remain as
the central problem in explosion research, both in terms of direct
applicability of its results to practical situations, and the
wealth of fundamental knowledge that can be deduced from its study.
It can be said that the qualitative understanding of the mechanisms involved is now fairly complete. However, little progress
has been made to date in terms of achieving a quantitative description of the phenomenon. Even empirical relationships of a
sufficient generality giving the flame acceleration rate or the
maximum flame speed attainable as a function of mixture combustion
and obstacle configurations have yet to be formulated. Even
though numerous experiments have been performed in the past years,
ranging from small laboratory sacle flame tubes of a few centimeters in diameter to very large tubes of 2.5 m diameter, and
flame geometries varying from planar to cylindrical in open fields
where the cloud covers an area of 600 m^2, quantitative information
regarding the dependence of the observed combustion processes on
the fluid dynamics field specific of the geometry and obstacle
used, have not been obtained. Most of the studies were made on
the initial transient development of the flame. It is this additional degree of complexity to an already extremely involved problem of non-linear coupling between fluid dynamics and chemistry
that prevents the results from being analyzed properly. The
author believes that special attention should be devoted instead
to the study of the steady state regime where a critical balance
between the positive and negative aspects of turbulence on combustion occurs. It is the accumulation of experimental data on this
steady state regime for the dependence of the flame speed on mixture composition, obstacle configurations and tube diameter that
will eventually permit the formulation of empirical correlations
leading to quantitative understanding. In the steady state regime,
experimental measurements of the very complicated turbulent flow
structure associated with the obstacle field are at least "inter-
pretable". This may then lead to more fundamental correlations
involving the use of fluid mechanics parameters rather than geo-
metrical parameters characteristic of the obstacle configuration.
Thus, correlations of a more universal nature emerge. In review-
ing the existing facilities of the various existing research
groups, there appears to be a lack in intermediate scale flame
tubes of the order of a meter in diameter and perhaps and L/D
ratio of the order of 100. It appears that, except for mixture
compositions well off-stoichiometry, the chemical length scales of
most fuels would not likely exceed the value where a strong dependence on tube diameters of about 1 m would still be present.
Thus, valuable data can be acquired readily in such intermediate
scale facilities. Although it may lack the "impressiveness" of a
half or full scale test of an actual geometry, the wealth of
scientific information obtained from these simple tube experiments

is well worth the effort. Little is known about the structure of
these steady state fast flames at present, even on a qualitative
basis. This should occupy some priority in a research program
since correct interpretation of any experimental results rest
heavily on having a clear picture of what the flame looks like.

Computer simulation using various codes should continue to
play an important role. However, the objectives must clearly be
defined. There is a big difference between the computer simulation of a certain experiment and the computer prediction of the
outcome of experiments as yet to be performed. Most of the existing codes are of the former nature and their usefulness rests on
their eventual development to become a predictive computer code.

The detonation phenomenon is in a much better position. Empirical correlations of sufficient unidersality already exist.
Together with simple analytical theories, practical information on
the dynamic detonation parameters can now be predicted with acceptable accuracy. However, all these correlations and theories
still hinge on the cell size being the fundamental input parameter.
Experimental determination of the cell size is not an easy task.
Smoked patterns are in general highly irregular, requiring an
"experienced eye" or a wealth of data to permit meaningful statistical averaging. There is little doubt that much effort is now
required to facilitate the determination of the cell size. At
present, the question of cell regularity and its dependence on
initial and boundary conditions (eg., ignition source, tube length
and geometry, etc.) is not clear. Fundamental research is needed
to support this quest for the development of a technique for
unambiguous cell measurements.

The problem of DDT has now emerged to be much less difficult
than it was once thought to be. In essence, the DDT problem reduces to the i) determination of the conditions necessary for the
onset of detonation, and ii) the fluid mechanics of turbulent
mixing of a reactive medium to achieve this set of critical conditions for the onset of detonation. Both problems can be investigated more or less in a decoupled fashion. The determination of
the conditions necessary for the onset of detonation is essentially a stability problem where one looks at the growth of shock
waves in medium at the verge of auto explosion. The second problem of mixing is a more complex one. There exists little data on
intense transient jet mixing. Clearly fundamental studies are
needed to acquire a prediction of the mixing rate for given initial
conditions.

Due to lack of systematic large scale DDT experiments at
present, it is judged important to carry some of these large scale
experiments to determine the critical dimensions of the mixing regions necessary for the various fuel-air mixtures.

Gas explosions will continue to be an active research area in
the foreseeable future as long as energy production and transport
play a dominant role in the economics of a country. However,
scientific progress can only be advanced rapidly if sufficient

communication and collaboration between various research institutions in the different countries can be effected. Unfortunately, self interests of sponsoring agencies often put severe restrictions on the open dissemination of information and discourages communication and collaboration of the scientists involved. It appears that this is perhaps the most important issue that has to be immediately resolved.

Acknowledgment

This paper was written almost half a year since the talk was given. Thus, the text resembles the talk only in the overall objective. Considerable amounts of new material have been added and I wish to thank Dr. Calvin Chan for providing me with the new results on fast turbulent flames in the third section of this paper.

Literature Cited

1. Lee, J.H., Moen, I. Prog. Energy Combust. Sci. 1980, 6, 359-389.
2. Lee, J.H., Guirao, C.M. Plant Operations Progress 1982, Vol. 2, No. 2, pp. 84-89.
3. Lee, J.H. Plant Operations Progress 1983, Vol. 2, No. 2, pp. 84-89.
4. Moen, I. "The Influence of Turbulence on Flame Propagation in Obstacle Environments" in "Fuel-Air Explosions"; Lee, J.H., Guirao, C.M., Eds.; University of Waterlook Press, 1982.
5. Lee, J.H. Fire Safety Journal 1983, Vol. 5, pp. 251-263.
6. Chapman, W.R., Wheeler, R.N.V. Journal Chem. Soc. 1926, 2139.
7. Shchelkhin, I.I. J. Exp. Theo. Phys. (USSR) 1940, 10, 823.
8. Wagner, H. "Some Experiments about Flame Acceleration", in "Fuel-Air Explosions", Lee, J.H., Guirao, C.M., Eds.; University of Waterloo Press, 1982.
9. Hjertager, B.H. "Numerical Simulation of Turbulent Flame and Reisure Development in Gas Explosions", in "Fuel-Air Explosions", Lee, J.H., Guirao, C.M., Eds.; University of Waterloo Press, 1982.
10. Geiger, W. "Quantification of Hazards from Vapour Cloud Explosions Research within the German Safety Program", in "Fuel-Air Explosions", Lee, J.H., Guirao, C.M., Eds.; University of Waterloo Press, 1982.
11. Zeeuwen, J.P., Van Wingerden Kees "On the Scaling of Vapour Cloud Explosion Experiments", paper presented at the 9th ICODGERS, Poitiers, France, 1983.
12. Urtiew, P.A. "Recent Flame Propagation Experiments at LLNL within the Liquified Gaseous Fuels Spill Safety Program", in "Fuel-Air Explosions", Lee, J.H., Guirao, C.M., Eds.; University of Waterloo Press, 1982.
13. Strehlow, R.A., University of Illinois, Champagne-Urbana, Private Communication (1983).

14. Berman, M., Sandia National Laboratory, New Mexico, Private Communication (1983).
15. Atkinson, R., Bull, D.C., Shuff, P.J. Combustion and Flame 1980, <u>39</u>, 297.
16. Meyer, J.W., Urtiew, P.A., Oppenheim, A.K. Combustion and Flame 1970, <u>14</u>, 13.
17. Thibault, P., Liu, Y.K., Chan, C., Lee, J.H., Knystautas, R., Guirao, C.M., Hjertager, B.H., Fuhre, K. Proc. 19th Symp. (Int.) on Combustion 1982, pp. 599-606.
18. Moen, I.O., Lee, J.H.S., Hjertager, B.H., Fuhre, K., Eckhoff, R.K. Combustion and Flame 1982, <u>47</u>, 31-52.
19. Hjertager, B.H., Combus. Sci. and Techn. 1982, <u>27</u>, 159-170.
20. Haselman, L.C. "TDC - A Computer Code for Calculating Chemically Reacting Hydrodynamic Flows in Two Dimensions", Lawrence Livermore National Laboratory, UCRL-52931, May 1980.
21. Hjertager, B.H. "Numerical Simulation of Turbulent Flame and Pressure Development in Gas Explosions", in "Fuel-Air Explosions", Lee, J.H., Guirao, C.M., Eds.; University of Waterloo Press, 1982.
22. Ashurst, W.T. and Barr, P.K. "Discolte Vortex Simulation of Flame Acceleration due to Obstacle Generated Flow", Sandia National Laboratory (Livermore) Report, SAND 82-8724 (1982).
23. Hjertager, B.H. Comb. Sc. Techn. 1982, <u>27</u>, 159-170.
24. Marx, K., Sandia National Laboratory (Livermore), Private Communication (1983).
25. Wolanski, P. "Gas and Dust Explosion Research in Poland", in "Fuel-Air Explosion", Lee, J.H., Guirao, C.M., Eds.; University of Waterloo Press, 1982.
26. Lee, J.H., Knystautas, R., Chan, C., Barr, P.K., Grcar, J.F., Ashurst, W.T. "Turbulent Flame Acceleration: Mechanisms and Computer Modeling", International Meeting on High-Water Reactor Severe Accident Evaluation, Cambridge, Mass., Aug. 28 - Sept. 1 (1983), also appeared as Sandia Report SAND 83-8655.
27. Taki, S., Fujiwara, T. Proc. 18th Symp. (Int.) on Comb. 1981, pp. 1671-1681.
28. Oran, E., Boris, J.P., Young, T., Flanigan, M., Burks, T., Picone, M. Proc. 18th Symp. (Int.) on Comb. 1983, pp. 1641-1649.
29. Bull, D.C., Elsworth, J.E., Shuff, P.J., Metcalfe, E. Comb. Flame 1982, <u>45</u>, 7-22.
30. Knystautas, R., Guirao, C.M., Lee, J.H.S., Sulmistras, A. "Measurements of Cell Size in Hydrocarbon-Air Mixtures and Predictions of Critical Tube Diameter Initiation Energy and Detonability Limits", 9th Int. Colloq. on Dynamics of Explosions and Reactive Systems (1983).
31. Knystautas, R., Lee, J.H., Guirao, C.M. Comb. Flame 1982, <u>48</u>, 63-83.
32. Edwards, D.H., Thomas, G.O., Nettleton, M.A. J. Fluid Mechanics 1979, <u>95:1</u>, 79-96.

33. Mitrofanov, V.V., Soloukhin, R.I., Soviet Phys.-Dokl. 1964, 9, 1055.
34. Lee, J.H., Knystautas, R., Guirao, C.M. "The Link Between Cell Size, Critical Tube Diameter Initiation Energy and Detonability Limits", in "Fuel-Air Explosions", Lee, J.H. Guirao, C.M., Eds.; University of Waterloo Press, 1982.
35. Dove, S.E., Wagner, Hg. Proc. 8th Symp. (Int.) on Comb. 1960, 589-600.
36. Donato, M. "The Influence of Confinement on the Propagation of Near Limit Detonation Waves", Ph.D. Thesis, McGill University, Montreal, 1982.
37. Vasiliev, A.A. Fiz. Goreniya Vzryva 1982, 18, 2, 132-136.
38. Lee, J.H.S. Ann. Rev. Fluid Mech. 1984, 16, 311-336.
39. Matsui, H., Lee, J.H. Proc. 17th Symp. (Int.) on Comb. 1978, 1269-1289.
40. Elsworth, J.E., Shell Research Limited, Thornton Research Center, U.K., Private Communication.
41. Benedick, W., Sandia National Laboratory, New Mexico, Private Communications (1983).
42. Knystautas, R., Lee, J.H., Moen, I., Wagner, Hg., Proc. 17th Symp. (Int.) on Comb. 1978, 1235-1245.
43. Pfförtner, I.C.T., Karlsruhe, West Germany, Private Communications.
44. Geiger, W., Battelle Institute, Frankfurt, W. Germany, Private Communications.
45. Moen, I., Defense Research Establishment Suffield, Alberta, Canada, Private Communications.

RECEIVED January 16, 1984

Chemical Kinetic–Fluid Dynamic Interactions in Detonations

ELAINE ORAN

Laboratory for Computational Physics, Naval Research Laboratory, Washington, DC 20375

We summarize a number of simulations aimed at deciphering some of the basic effects which arise from the interaction of chemical kinetics and fluid dynamics in the ignition and propagation of detonations in gas phase materials. The studies presented have used one- and two-dimensional numerical models which couple a description of the fluid dynamics to descriptions of the detailed chemical kinetics and physical diffusion processes. We briefly describe, in order of complexity, a) chemical-acoustic coupling, b) hot spot formation, ignition and the shock-to-detonation transition, c) kinetic factors in detonation cell sizes, and d) flame acceleration and the transition to turbulence.

In this paper we describe some of the basic effects which arise specifically from the coupling between chemical kinetics and fluid dynamics in combustion systems. Although the particular emphasis here is on the ignition and propagation of detonations, many of the more fundamental interactions described are generally applicable to flames. The selection of topics is by no means meant to be comprehensive; rather, it represents a potpourri of ideas which complement each other and those presented in the other papers in this session of the Symposium. Although we mainly use calculations performed at NRL to extract and illustrate the details of the interactions, we have drawn liberally on the results of experiments and analytic theory.

Chemical-Acoustic Coupling

Studies of chemical-acoustic coupling are concerned with the interactions between sound waves and the processes involved in chemical reactions, which could involve either the generation of new species or energy release or absorption (1). There are essentially two approaches to discussing chemical-acoustic inter-actions. The first is to consider the effects of energy release or other nonequilibrium properties of the mixture on the sound waves themselves. The earliest work in this area was done by Lord Rayleigh in the nineteenth century. Figure 1 shows a summary of Lord Rayleigh's results on the effects of energy release on the phase and amplitude of sound waves. We see that exactly how the frequency and amplitude of the sound wave are altered depend on where in the cycle of the sound wave the energy is released. Since this early work, significant theoretical results have been obtained which describe how sound waves attenuate as they travel through mixtures slightly perturbed from chemical equilibrium (2-4). We know that sound waves can be amplified as they propagate through nonequilibrium mixtures (5,6) and that a shock forms before a thermal explosion when a sound wave propagates through an explosive mixture (7,8). Finally, we have learned that sound waves propagating through exothermic mixtures can be amplified and new modes can be excited (9,10). In particular, sound waves are amplified during the process of chemical energy release itself, and this effect has a tendency to die out once the system has reached a new equilibrium

The other approach to this problem is to look at how the presence of the sound wave changes the chemical reaction process. First, we know that fluctuations in the temperature and pressure can alter chemical reaction times. Gilbert et al. (11) have shown how single-step unimolecular dissociation or conversion can produce a predictable amplitude and dispersion change in a sound wave. From this they proposed using acoustic changes to help follow and diagnose chemical reactions. Toong and coworkers (9,10) have shown that sound waves enhance energy release rates in exothermic systems.

All of the analysis described in the last two paragraphs has been done on systems with at most an idealized, one-step Arrhenius kinetics model, and most of the analysis has been done in a linear approximation. Recent work by Oran and Boris (12) has used numerical simulations to show that sound waves can not only affect the energy release time as Toong has shown, but can also drasti-cally reduce the the period before there is any energy release, which we define as the chemical induction time. These simulations coupled a detailed chemical reaction mechanism for hydrogen-oxygen combustion to a solution of the conservation equations for mass, momentum and energy. The shortened induction time is a property of the nonlinear coupling between the intermediate species' con-centrations and the changes in density and temperature which result form the presence of the sound waves. For this effect to occur, it is necessary to have a number of species and a number of

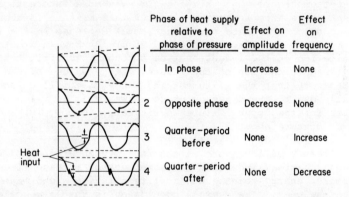

Figure 1.   Rayleigh's criterion for the effects of heat release on the amplitude and phase of a sound wave.

reaction   rates coupling them: it is not a result that would come
from an analysis which was based on a one-step Arrhenius reaction
rate.

   Figure 2 shows two calculations which have been done with
the detailed model described above. The simulations are for
hydrogen-oxygen mixtures diluted with argon. For both cases the
period of the imposed sound wave was chosen so that there are
about three periods within a chemical induction time. The amplitude of the imposed sound wave was the same in both cases. Each
figure shows the temperature versus time profiles at three locations in the mixture. In the upper figure, there is about a fif-
teen microsecond difference in the induction time generated by the
presence of the sound wave. There is a 150 microsecond difference
in the bottom figure, but this case is even more interesting if we
note that the static induction time for this mixture is about 1500
microseconds. Thus there has been a total decrease in the induction time of about a factor of ten. Figure 3 shows a static
quantification of the induction time through a sensitivity parameter defined on the top of the figure. Large values of the contours indicate that the mixtures' induction time will change substantially due to the presence of sound waves. The value of this
parameter for the upper mixture in Figure 2 is about 10, and for
the lower mixture about 35.

   The result of this work is that the presence of sound waves
can change chemical induction times and energy release times.
Furthermore, energy release can amplify existing sound waves and
generate new ones. The overall conclusion is that timescale
analyses may have to be modified to account for the presence of
these coupling effects.

Hot Spots, Reactive Centers, and the Shock-to-Detonation Transition

Local fluctuations in temperature, density and pressure are always
present, even in homogeneous, premixed systems. In systems whose
sensitivity to perturbations is high, these fluctuations can have
surprising consequences and alter the behavior from conventional
expectations. We saw this above when we investigated the sensitivity of the hydrogen-oxygen mixture to the presence of sound
waves. It has also been shown that the same systems are just as
sensitive to entropy perturbations (12). Below we describe a
system in which the combined effects of sound wave and entropy
perturbations cause non-ideal behavior behind incident shocks.
The study is based on the recent shock tube studies by Edwards
et al. (13) which have shown ignition starting from one of a
series of hot spots behind the shock.

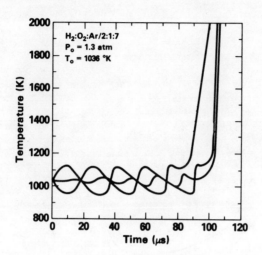

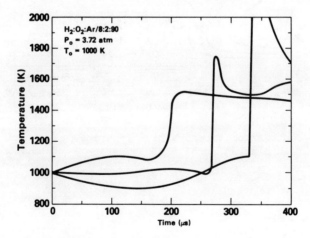

Figure 2. Calculated temperature vs. time at three locations in reactive mixtures that are perturbed by a sound wave.

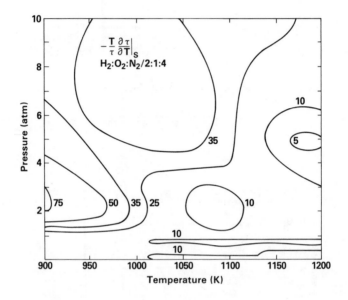

Figure 3.   Contours of the sensitivity parameter defined at the top of the figure as a function of temperature, T, and the induction parameter, τ.

In order to explore possible mechanisms and understand the implications for sensitive systems, we have performed a series of numerical simulations using the model described in the last section of this paper (14). Specifically, the calculations simulated an incident shock driven by a high pressure helium driver into a low pressure region containing a combustible mixture of hydrogen in air. Figure 4 shows the results from one such calculation which shows a rather surprising effect: ignition does not occur at the contact discontinuity, which is the region which is heated for the longest time. A careful look at these calculations shows that some small amount of energy release has originally occured at the contact discontinuity, and this has generated pressure pulses which have traveled forward and accelerated the shock front. This process has created a region behind the shock front which is at a higher temperature and pressure than that of the original shock front. Since the mixture is in a sensitive region of the temperature-pressure plane, the induction time behind the shock now has been somewhat reduced. This process repeats itself until one of the reaction centers ignites and forms a reaction wave. Since the calculation is one-dimensional and Cartesian, there are really two reaction waves generated: one moving forward in the direction of the incident shock, and another moving backwards toward the contact discontinuity. This is shown in Figure 5. The forward moving wave can transition to a detonation even before it reaches the incident shock wave. When it reaches the shock wave, there is an abrupt increase in shock velocity. The backward moving wave travels more slowly and when it reaches the contact discontinuity, it sends a pressure pulse into the driver gas. Figure 6 shows the time history of the development of the reaction centers and the process of ignition and transition to detonation.

Thus we have seen another case where the interaction of sound waves and entropy perturbations with chemical reactions has altered the timescales and even the location of the physical processes. This is a much more complicated and less idealized example than the sound wave study described in the previous section. The sound wave calculations, however, did isolate the interactions and looked at what was required to quantify the effect. Here we have not only seen the direct effect of variations in the induction times due to the presence of perturbations, but also the indirect effect of an alteration of the background physical conditions, i.e., the Mach number of the shock, due to pressure pulses generated by energy release.

Kinetic Factors and Detonation Cell Sizes

It has been known for some time that a smooth, planar propagating detonation is unstable. This means that small perturbations can trigger the instability and the system may change to a more stable structure. For example, Erpenbeck (15) developed a criterion for

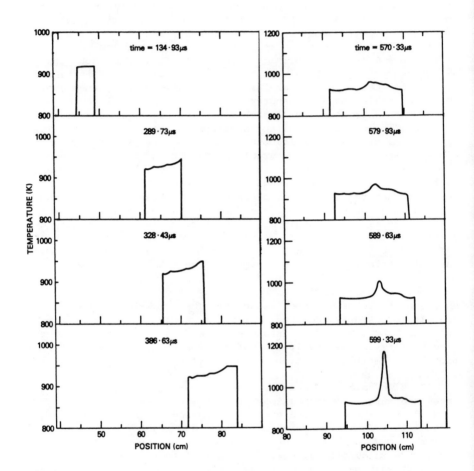

Figure 4.   The upper portions of calculated profiles of temperature as a function of position from a simulation of a shock propagating in a reactive hydrogen-oxygen mixture that has a relatively
high sensitivity as determined from Figure 3.

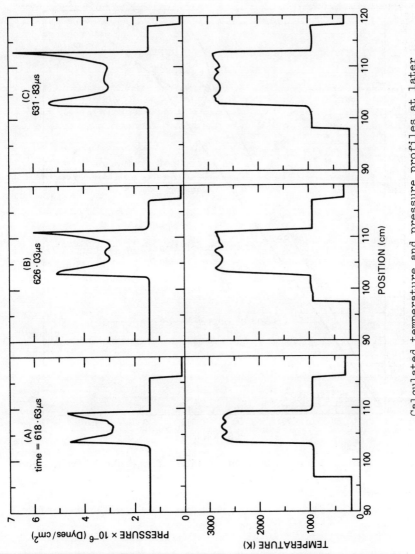

Calculated temperature and pressure profiles at later times than in Figure 4.

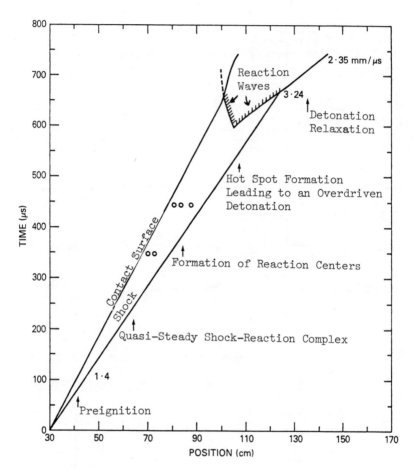

Figure 6.   Location as a function of time of the shock front, contact surface reaction centers, reaction waves, and detonation front for the calculation shown in Figures 4 and 5.

when an overdriven detonation front was stable, given an activation energy and a single-step Arrhenius reaction rate. Such a
stability analysis gives a "go, no-go" answer, but it does not
give information about the evolution of the instability. This
aspect was treated by Fickett and Wood (16), who performed numerical simulations of Erpenbeck's model problem. They did indeed
find that the front oscillates due to the coupling of the fluid
dynamics and energy release.

We also know now that detonations do not propagate as smooth
fronts. Their equilibrium structure is composed of a set of
interacting, intersecting shock waves, which we call the incident
shock, the Mach stem, and transverse waves. Further, we know that
the intersection of the incident shock and Mach stem is a triple
point whose position in time describes the rhombic shape we call a
"detonation cell." The incident shock is not steady, but is continuously decaying; and the transverse wave, a reflected shock
intersecting the Mach stem and the incident shock, shuttles back
and forth across the detonation front. A detonation cell is
reinitiated when collisions occur between transverse waves moving
in opposite directions. Behind the shock fronts, there is a reaction zone which varies in distance from the front depending on the
length of time it has been since cell reinitiation. The velocity
of the leading shock varies from above the Chapman-Jouguet value
to below it and takes a sudden jump when the cell structure is
reinitiated by transverse wave collisions. Detonation cells have
been measured by coating the inside of a detonation tube with soot
and letting the triple point trace out the pattern. In general
the pattern is quite irregular, but a characteristic cell size can
be determined for a particular material at a given temperature and
pressure. Excellent summaries of what is known about the cell
structure have been given by Strehlow (17) and by Fickett and
Davis (18).

Thus we see that the factors that determine the detonation
cell size are complicated interactions of fluid dynamics and
chemical kinetics. The fluid dynamics here involves a number of
interacting shock waves, pressure waves, sound waves, and perturbations due to energy release. The chemical reactions are occurring in an environment which is always subjected to fluctuations
and pressure perturbations. Thus we have gone up in level of
complexity in the flow properties. However, this problem encom-
passes all of the issues we discussed in the previous section.

Numerical simulations of the structure of multidimensional
detonations have been carried out by Taki and Fujiwara (19) and by
Oran et al. (20). Because of the expense involved in these simulations, both groups used a phenomenological model meant to
describe the basic features of energy release, and coupled this to
a solution of the conservation equations for the fluid dynamics.

Below we summarize some of the results obtained by Oran et al.
(20), which used chemical models that had the ability to represent
the change in chemistry due to pressure and temperature perturbations in the chemically sensitive regime.

As mentioned above, the cell size of a detonation depends on
the particular mixture and its pressure and temperature. Thus the
number of detonation cells that will be formed by a self-propaga-
ting detonation in a tube depends on the height of the tube. In
practice, however, if the tube is not high enough, the detonation
will extinguish due to losses to the walls. For the calculations,
however, we do not have to include wall losses, and we assume for
now that we are discussing two-dimensional Cartesian geometry.
The experimental equivalent would be a very thin tube. If the
tube height is less than one cell height, a perturbed half-cell
structure will evolve. If the tube height is one cell height, a
full cell will develop, etc. Note that because of the symmetry of
the cells, it should be possible to obtain many of the two-dimensional features by simulating a half of a cell.

Figure 7 shows the results of simulations of a detonation
propagating down a tube filled with a mixture of hydrogen, oxygen
and argon. The tube height is slightly greater than half of a
detonation cell height. The simulation was done for a marginal
detonation: that is, a detonation which is close to the values of
pressure, temperature, and stoichiometry for which it will die
out. Figure 7 contours a quantity we call the induction parameter, which is a measure of the amount of material reacted.
Superimposed on these contours is the shock front as determined by
the jump of temperature and pressure. Thus the material to the
right is totally unshocked and unreacted. The material to the
left of the temperature contour is in various stages of reaction
completion.

The interesting feature which is noticeable both in these
simulations and in the experiments by Edwards (20) is the formation of the pockets of unburned gas which occurs when the transverse wave collides with the wall. Because of the symmetry of a
cell, this is equivalent to the collision of two transverse waves.
Figure 8 shows contours from a simulation in a much shorter tube.
The advantage of this is that the half cell structure is forced
and the calculation is much more resolved. Here it is much easier
to see the pockets being formed. It is particularly interesting
to note that the pockets have a temperature and pressure that are
in an extremely sensitive regime in the temperature-pressure
plane, as determined from calculations such as those in Figure 3.
Thus many of their burning properties and their energy release
properties will be determined not by their initial pressure and
temperature, but by the myriad of pressure and entropy

perturbations which they feel due to the extremely dynamic
environment in which they exist.

There are also a number of implications of these unreacted
gas pockets. The first is that we now have a way of creating an
inhomogeneous material out of a homogeneous material. The pockets
of material are at different pressures and temperatures, and have
a different composition than the completely burned material around
them. Second, such pockets could provide the perturbation necessary to initiate an instability and thus provide the initial impe-
tus to a mechanism that would allow the material to form a new
number of detonation cells characteristic of the chamber size. In
an open environment where there are no walls, the pockets could
provide the perturbation necessary to allow new cells to generate
and the number of cells to increase. And finally, the pockets
could provide a mechanism for the extinction of detonations. A
scenario could evolve in which more and more energy release is
delayed because the pockets become bigger and bigger.

Flame Acceleration and Transition to Turbulence

Here we are interested in mechanisms of the transition process.
Some of the basic questions we must address are fluid dynamics
questions: how do laminar flows transition to turbulent flows and
what are the mechanisms of vorticity generation. Then we ask how
the presence of chemical reactions and energy release alter these
situations or generate additional mechanisms through coupling
interactions. This involves questions such as how do flames
stretch and increase their surface area, and thereby increase the
burn rate and flame velocity.

The transition process is often initiated by one of a number
of instabilities such as a Rayleigh-Taylor instability, in which a
heavy fluid is accelerated through a light fluid, or a Kelvin-
Helmholz instability which is the classic shear flow instability.
Figure 9 shows a two-dimensional simulation of a Rayleigh-Taylor
instability which was done with a Lagrangian numerical method in
which the usual quadrilateral computational grid is replaced by a
grid of triangles (21). Such an instability at a propagating
interface provides a mechanism for flame stretching and thus flame
acceleration. Figure 10 shows a two-dimensional FCT Eulerian
simulation of a Kelvin-Helmholz instability generated at the
interface between two coflowing gases of different velocities
(22). The coherent structures which form through this interaction
are important for mixing and the transition to turbulence in gas
jets.

There is also another method of generating vorticity and thus
a way of transition to turbulence which is not based on a fluid
instability, but simply on the straightforward generation of

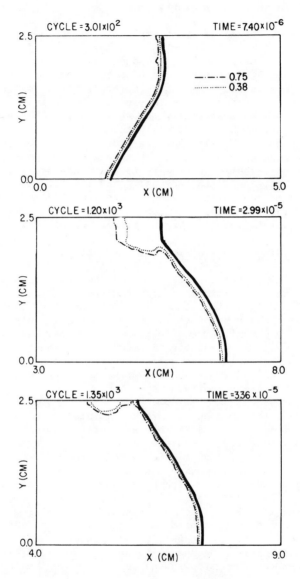

Figure 7a. Contours of the induction parameter for calculations of a detonation propagating in a mixture of hydrogen, oxygen, and argon. High values of the induction parameter indicate that the material is about to release energy. The position of the shock front is marked by the heavy dark solid line.

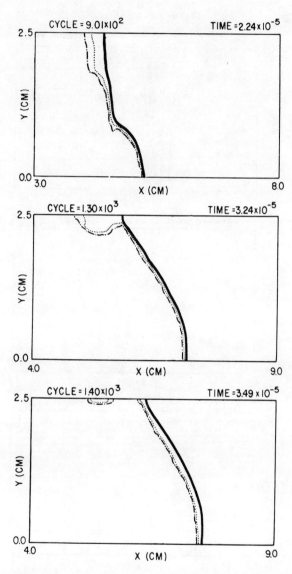

Figure 7b. Contours of the induction parameter for calculations of a detonation propagating in a mixture of hydrogen, oxygen, and argon. High values of the induction parameter indicate that the material is about to release energy. The position of the shock front is marked by the heavy dark solid line.

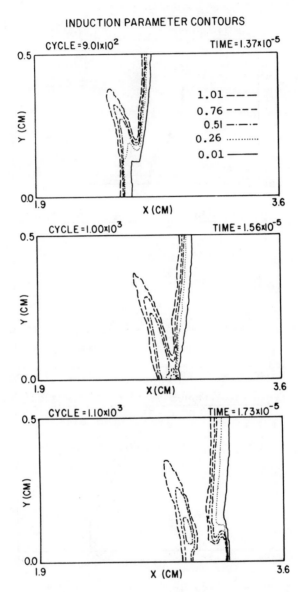

Figure 8a.   Same as Figures 7a and 7b except for a smaller tube.

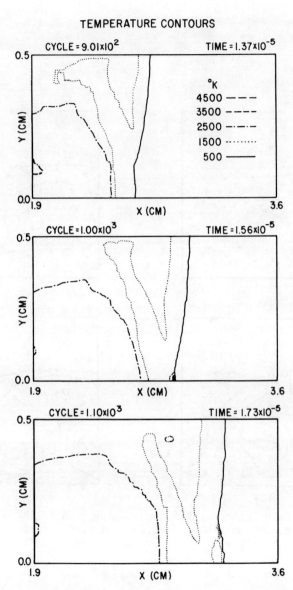

Figure 8b. Same as Figures 7a and 7b except for a smaller tube.

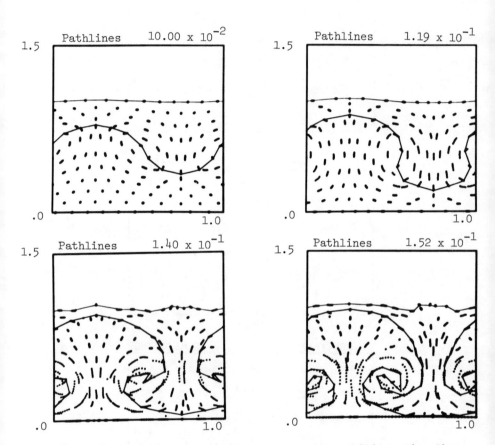

Figure 9. Calculation of Rayleigh-Taylor instability using the Lagrangian technique with automatic zone restructuring. A heavy fluid falls through a light fluid, and there is a free surface on the top.

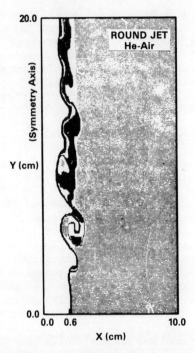

Figure 10. One frame from the calculation of the evolution of a
Kelvin-Helmholtz instability at the surface of a round jet of air
into an air background. Cylindrical symmetry was used and the axis
of symmetry is the left-hand boundary. The inflow speed is about
10^4 cm/s, and the outer co-flow is about 10^2 cm/s.

vorticity by the interaction of pressure waves with density gradients or obstacles. These are two quite different mechanisms. The
interaction with density gradients has recently been discussed by
Picone et al. (23) who look at vorticity generation through the
interaction of weak shocks and density gradients typical of an
expanding flame. Through the interaction of the pressure wave
with the flame, the scale of the inhomogeneity is halved. Then as
another pressure wave interacts with this newly created inhomogeneity, the scale is again halved. Figure 11 shows results of
Picone's which were done to simulate shock tube experiments by
Markstein (24). As the planar shock passes through the spherical
density perturbation, vorticity is generated and the scale of the
inhomogeneity is halved. As the reflected shock passes through
the density gradient again, the scale is halved again.
Markstein's experiments were done with a flame ignited in a shock
tube. Picone's calculations were done for a density discontinuity
modeled on that expected for a flame.

Conclusions and Summary

In the examples given above we have tried to describe some of the
phenomena which arise as a result of chemical kinetic-fluid
dynamic coupling. First, we described studies of the isolated
effects of chemical-acoustic coupling, emphasizing the effects on
the chemical kinetics. The major conclusion is that sound waves
and entropy perturbations can alter chemical timescales, and that
this effect can be quantified. We then described a system in
which sound waves and entropy perturbations behind a shock wave
caused early ignition at unpredictable locations and at reduced
ignition times. A series of reaction centers formed and one of
these close to the shock front eventually ignited.

The last two sections dealt with systems in which the flow
was two-dimensional and thus substantially more complicated.
First we described some of the properties occurring in propagating
detonations, for which the structure is highly dependent on the
chemical kinetic-fluid dynamic interactions. Finally, in the last
section we discussed some processes and mechanisms involved in the
transition to turbulence, which is important for flames.

As we proceeded into the various sections in this paper, we
described a series of progressively more complex interactions.
Each new topic contained most of the complications present in the
previously discussed topics. The basic ingredients in the studies
presented are reasonable representations of both the flow field
and the chemical kinetics. The complexity arose because we considered multidimensional effects and expanded the range of scales
considered. In general, the current level of understanding of the
details of the interactions between the chemistry and flow is less
accurate as the flow and the chemical reactions become more and

DENSITY CONTOURS

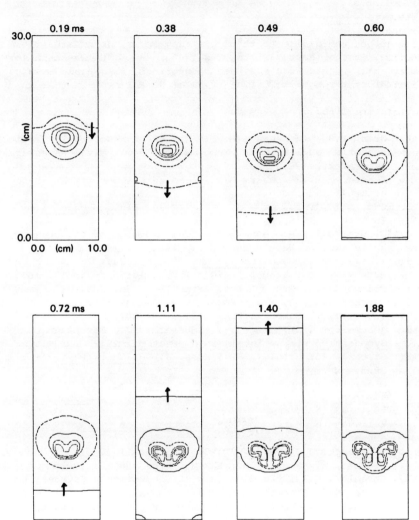

Figure 11.   Density contours for the simulation of a weak shock
passing through a density gradient.   Time is marked on top of each
frame.   The incident shock is passing through the gradient at 0.19
ms, and by 1.11 ms the reflected shock has passed through.

more complex. However, as we have tried to show, we are just now
at the stage where we can begin to include this effect accurately
in simulations.

Acknowledgments

The author would like to thank Drs. Jay Boris, John Gardner,
Michael Picone, Marty Fritts, Ted Young, and K. Kailasanath for
their help. This work has been sponsored by the Naval Research
Laboratory through the Office of Naval Research.

Literature Cited

1. Oran, E.S.; Gardner, J.H., "A Review of Research in Chemical-
    Acoustic Coupling," NRL Memo. Report 5121, Naval Research
    Laboratory, 1983.
2. Einstein, A., Sitzber. Deut. Akad. Wiss. Berlin, Kl.-Math.-
    naturn., 1920, 380.
3. Chu, B.T., Proc. of Heat Transfer and Fluid Mechanics
    Institute, 1958, p. 80.
4. Clarke, J.F.; McChesney, M., "The Dynamics of Real Gases";
    Buttersworth: Washington, D.C., 1964, p. 182.
5. Clarke, J.F., Comb. Sci. Tech. 1973, 7, 241.
6. Srinivasen J.; Vincenti W.G., Phys. Fluids 1975, 18, 1670.
7. Clarke, J.F., Prog. Astro. Aero. 1981, 76, 383.
8. Blythe, P.A., J. Fluid Mech. 1969, 37, 31.
9. Toong, T.Y., Combust. Flame 1972, 18, 207.
10. Abouseif, G.E.; Toong, T.Y.; Converti, J., Seventeenth
    Symposium (International) on Combustion, 1979, p.1341.
11. Gilbert, R.; Ortoleva, P.; Ross, J., J. Chem. Phys. 1973, 58,
    3625.
12. Oran, E.S.; Boris, J., Combust. Flame 1982, 48, 149.
13. Edwards, D.H.; Thomas, G.O.; Williams, T.L., Combust. Flame
    1981, 43, 187.
14. Kailasanath, K.; Oran, E.S., Combust. Sci. Tech. 1983, to
    appear.
15. Erpenbeck, J.J., Phys. Fluids 1965, 8, 1192.
16. Fickett, W.; Wood, W.W., Phys. Fluids 1966, 9, 903.
17. Strehlow, R.A., "Fundamentals of Combustion", Krieger: New
    York, 1979; Chapt. 9.
18. Fickett, W., and Davis, W.C., "Detonation", University of
    California Press: Berkeley, 1979.
19. Taki. S.; Fujiwara, T., Eighteenth Symposium (International)
    on Combustion, 1981, p. 1671.
20. Oran, E.S.; Young, T.R.; Boris, J.P.; Picone, J.M.; Edwards,
    D.H., Nineteenth Symposium (International) on Combustion,
    1982, p. 573.
21. Fritts, M.J.; Boris, J.P., J. Comp. Phys. 1979, 31, 173.

22.   Boris, J.P.; Fritts, M.J.; Oran, E.S., "Numerical Simulations of Shear Flows in the Splitter Plate and Round Jet", Naval Research Laboratory Memo Report, to appear 1983.
23.   Picone, J.M.; Oran, E.S.; Boris, J.P.; Young, T.R., Proc. 9th ICODERS, To appear 1984.
24.   Markstein, G.H., "Nonsteady Flame Propagation," Pergamon: Oxford, 1964.

RECEIVED October 28, 1983

Chemical Kinetic Factors in Gaseous Detonations

CHARLES K. WESTBROOK

University of California, Lawrence Livermore National Laboratory, Livermore, CA 94550

Computer modeling techniques have been applied to
the study of hydrogen and hydrocarbon oxidation in
gaseous detonation waves. Characteristic reaction
times and lengths are computed which correlate well
with observed detonation parameters, including
critical tube diameters for transition to spherical
detonation, detonation cell sizes, critical initiation energies, and lean and rich limits for detonation in a linear tube. Inhibition or extinction of
a detonation is shown to occur from increases in
the ignition delay time, and increased detonability
or kinetic sensitization results from decreased
ignition delay times.

Detonation waves are an important class of combustion phenomena,
due both to the potential safety hazards which they represent
and to the insights into fundamental combustion processes which
they provide. Gaseous detonations have been examined for many
years, in both experimental and theoretical studies. More recently, computer modeling studies of detonation waves have begun
to appear. The chemical kinetics submodels have been considered
to be the weakest part of existing detonation models. However,
recent development of comprehensive kinetic reaction mechanisms
for the oxidation of many practical fuels (1,2) has changed this
situation significantly.

The present paper reports progress that has been made on
chemical kinetics in detonations and how well kinetic predictions correlate with available experimental data. The success
of this approach in reproducing experimental data illustrates
the central role of kinetics in detonations, and it suggests
strongly that this technique provides a reliable basis for pre-
dicting detonation properties for conditions which have not yet
been explored experimentally. For example, very few experimental data are available for detonation properties at initial

pressures above 1 atm, at initial temperatures different from normal room temperature, or for mixtures in which the oxidizer is air rather than oxygen. All these can be of extreme practical interest to the industrial community in helping to understand the hazards associated with explosive mixtures and to know how accidental disasters can be prevented. As a further extension of this type of approach, this type of modeling can suggest kinetic means of modifying the detonation parameters of a given fuel-oxidizer mixture, either enhancing or inhibiting detonability through the use of appropriate chemical additives.

Chemical Kinetics

At the present time, the fuels which can be described by this modeling approach include hydrogen, carbon monoxide, methane, methanol, ethane, ethylene, acetylene, propane, and propylene. The reaction mechanism used to describe the oxidation of these fuels has been developed and validated in a series of papers (3-7). The elementary reactions and their rate expressions are summarized in Reference (7) and are not reproduced here due to space limitations. Reverse reaction rates are computed from the forward rates and the appropriate thermodynamic data (8). This mechanism has been shown to describe the oxidation of methane (3,4), methanol (5), ethylene (6), and propane and propylene (7) over wide ranges of experimental conditions. It has also been used to describe the shock tube oxidation of ethane (4,9), and acetylene (10).

The parameter regimes in a detonation are similar to those in shock tubes, so the most important test of this type of mechanism is its ability to reproduce shock tube ignition data. One example of this validation process compared computed ignition delay times (7) with experimental results of Burcat et al. (11). In the experiments, mixtures of propane, oxygen, and argon were studied in reflected shock waves at initial temperatures from 1250 to 1700 K, pressures from 2 to 15 atmospheres, and equivalence ratios from 0.5 to 2.0. From the experimental results, it was found that the ignition delay time τ could be approximated in terms of the initial temperature T_0 and reactant concentrations (in moles/cm^3) by

$$\tau = 4.4 \times 10^{-14} \exp(21240/T_0) \; [C_3H_8]^{0.57}[O_2]^{-1.22} \text{ sec.}$$

In each computation τ was defined as the time corresponding to the maximum rate of reaction between CO and O atoms. Other realistic definitions of τ, such as the time of maximum rate of pressure or temperature rise, gave nearly identical results. From the computed ignition delay times and initial reactant concentrations, model values of the correlation function β

$$\beta = \tau \; [O_2]^{1.22} \; [C_3H_8]^{-0.57} \; \mu s \; (\text{mole/cm}^3)^{0.65}$$

were calculated, and the results are summarized in Figure 1.
The solid line represents the overall experimental correlation
function of Burcat et al., while the individual symbols represent computed values of β. The specific conditions for each
Mixture identified in Figure 1 can be found in Reference (11).
The general agreement between computed and experimental data is
very good, indicating that the reaction mechanism is properly
reproducing the shock tube ignition behavior for this fuel.
Similar detailed comparisons have been carried out in previous
modeling studies with basically the same reaction mechanism for
ethylene (6), methane and ethane (4), and methanol (5). The
H_2 oxidation submechanism has been extensively validated, with
nearly all of the elementary reaction rates being well known.

In the past, detonation models have used global rate expressions to compute chemical induction times for fuel-oxidizer mixtures, but such expressions are often not satisfactory, even
when they have been based on shock tube data. Most shock tube
experiments are carried out with high dilution by Ar, He, or
N_2, so that fuel and oxygen concentrations are usually quite
low. However, overall reaction order and activation energy in
global induction time correlations often change with the amount
of dilution. Global rate parameters can also change with equivalence ratio, pressure and temperature. As a result, induction
times computed from global expressions can be seriously in error
when applied to undiluted fuel-oxidizer mixtures, making a detailed kinetic mechanism an essential part of the present detonation model.

The kinetic submechanism for the inhibition studies was also
developed by a sequential process, beginning with HBr (12) and
the other halogen acids HCl and HI, followed by reactions involving methyl, vinyl, and ethyl halides (13) and CF_3Br (14).
The inhibition mechanism and reaction rates are given in
Reference (13).

Detonation Model

The model used here is the Zeldovich-von Neumann-Doring (ZND)
model in which, locally, a detonation consists of a shock wave
traveling at the Chapman-Jouguet (CJ) velocity, followed by a
reaction zone. The shock wave compresses and heats the fuel-
oxidizer mixture which then begins to react. In most mixtures
the fuel oxidation consists of a relatively long induction
period during which the temperature and pressure remain nearly
constant, followed by a rapid release of chemical energy and
temperature increase. For each fuel-oxidizer mixture, a calculation is first made of the relevant CJ conditions. From the
detonation velocity D_{CJ}, the conditions in the von Neumann
spike, including the temperature T_1, pressure P_1, and
particle velocity u_1 of the post-shock unreacted gases can be
computed and then used as initial conditions for the chemical

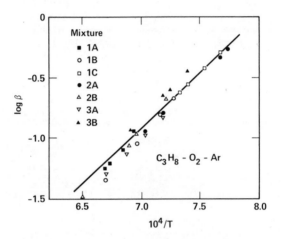

Figure 1.   Correlation functions for shock tube ignition of propane.   Solid line is overall correlation from Burcat et al. ($\underline{11}$);
symbols are computed values.

kinetics model. The shock velocity varies within a single detonation cell from an initial value of about 1.6 D_{CJ} to a minimum of about 0.6 D_{CJ}, so the CJ conditions used here represent average values, and computed induction times will also be averages.

The reactive mixture is assumed to remain at a constant volume over its reaction time, and the induction time is defined in terms of its temperature history. Most of the mixtures underwent a large temperature increase of more than 1000 K, and the induction time is defined as the time of maximum rate of temperature increase. In most cases, this coincides approximately with the time at which the temperature has completed about half of its total increase. This is not, strictly speaking, a true induction period, often defined as the time required for a small (i.e. 1-5%) temperature or pressure increase, but it represents a time scale for the release of a significant amount of energy. In addition to the induction time τ, it is useful to define the induction length $\Delta \equiv \tau(D_{CJ}-u_1)$, which represents a characteristic length scale in the post-shock unreacted gas mixture.

As a result of these simplifications, the computed induction times and lengths define characteristic time and length scales rather than the precise history of a gas element through the detonation front. The evolution of the reacted gas subsequent to the induction period considered here is dominated by the fluid mechanics of the post-induction expansion of the reaction products. This expansion reduces the pressure and density of these products and alters the kinetic equilibrium, leading eventually to the CJ state. Since virtually all of the reactants have been consumed by this time, the kinetics of this final expansion phase are controlled by relatively slow radical recombination processes. The present model does not attempt to follow that entire relaxation phase, concentrating on the details of the induction kinetics in the von Neumann spike.

This model of the detonation neglects some potentially significant effects arising from hydrodynamic-kinetic interactions. Variations of density, temperature, and particle velocity in the post-shock unreacted mixture are not considered. Multiple shock wave reflections, rarefactions, interactions with confining walls, cellular structure, and related effects are also not treated directly by the present simplified model. A really comprehensive detonation simulation model will have to include such interactions and at least two and probably even three space dimensions, but such a treatment is beyond the scope of the present formulation. In cases where such interactions are important, this approach may not be adequate.

Results

Many CJ and induction delay time calculations have been carried

out, examining the effects of variations in many physical and
chemical parameters. For fuel-O_2 and fuel-air mixtures, the
equivalence ratio can be varied over the complete range from
pure oxidizer to pure fuel. For stoichiometric fuel-O_2 mixtures the amount of dilution by N_2 can be varied between β =
0 (β is the initial ratio of N_2 to O_2, so β = 0 corresponds to a fuel-O_2 mixture) and β = 3.76 (i.e. fuel-air).
Other diluents such as CO_2, H_2O, and Ar have been used. The
initial pressure has been varied between 0.01 and 10.0 atmospheres, and the initial temperature between 200 K and 500 K in
stoichiometric fuel-O_2 and fuel-air. Finally, small amounts
of selected halogenated species have been included in fuel-air
mixtures initially at atmospheric temperature and pressure.

As an example, computed values of the induction length Δ
are summarized in Figure 2, showing the effects of variations in
the fuel-oxidizer equivalence ratio ϕ. The dashed curves
represent fuel-air and the solid curves fuel-O_2 mixtures.
Results for propane (not shown) are very similar to those for
ethane. All of these results assume that the initial pressure
is atmospheric and the initial temperature is 300 K. There is a
significant difference between the hydrocarbon results and those
for hydrogen. For each hydrocarbon, values of Δ for fuel-O_2
are lower than those for fuel-air by a factor of about 100, and
the minimum value of Δ falls on the rich side of stoichiometric. However, for H_2-O_2 mixtures, values of Δ are only
a factor of 10 lower than those for H_2-air (15), and the minimum value of Δ occurs on the lean side of stoichiometric. Induction lengths have been computed for a wide variety of other
initial conditions as well, and the remaining sections of this
paper will show how these induction lengths can be related to
specific detonation properties.

Detonation limits. The most important result of the kinetic
modeling work has been the observation that the induction length
Δ is approximately proportional to the characteristic detonation cell width a. For stoichiometric fuel-air and fuel-oxygen
mixtures initially at atmospheric temperature and pressure, the
ratio a/Δ is approximately 20 and 35 respectively, based on
observed cell sizes from experimental studies (16-18). Similar
agreement is obtained between computed induction lengths and
experimental cell size data at initial pressures that are different from atmospheric (19,20). Correlations with cell size
data of Manzhalei et al. (19) for CH_4-O_2, C_2H_2-O_2, and
H_2-O_2 mixtures are shown in Figure 3. Here initial pressures are varied between 0.01 and 10 atmospheres, and the agreement with computed results is good when a/Δ = 29, consistent
with the proportionality obtained above at atmospheric pressure. The computed results reproduce even the slight curvature
in the H_2-O_2 measurements. This proportionality can then be
used to predict the lean and rich limits of detonation in linear

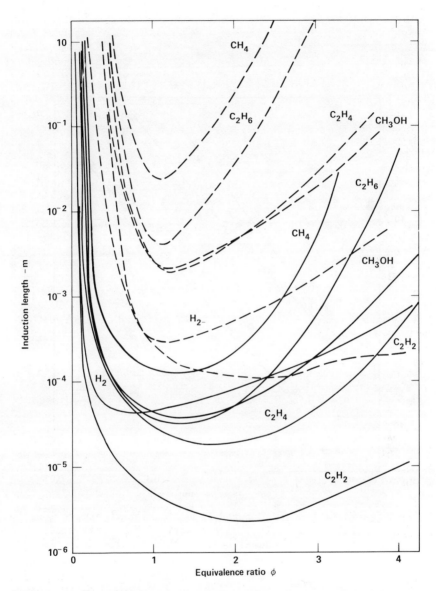

Figure 2.   Computed induction lengths for fuel-oxidizer mixtures. Dashed curves are for fuel-air mixtures; solid curves for fuel-O_2 mixtures.

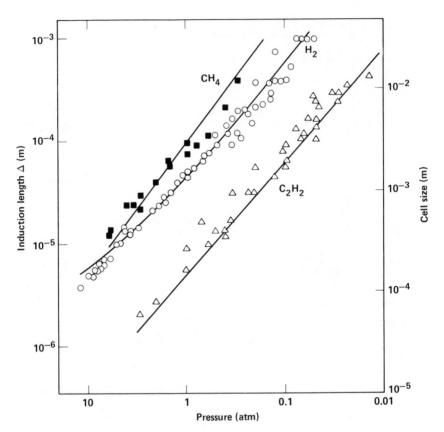

Figure 3. Variation of computed induction length and cell width
a with initial pressure in stoichiometric fuel-O_2 mixtures. Open
symbols indicate experimental cell width data from Manzhalei et
al. (19).

tubes, if it is assumed that single-head spinning detonation occurs at the limits with a transverse wave spacing that is twice the tube diameter. This has been done (15,21-24), using experimental data from a variety of sources (25-29). Using the computed induction length, the experimental tube diameter can be used to predict the limit compositions. Alternatively, the limit composition can be given and the model will predict the tube diameter required.

Critical tube diameter. It has been found (30-32) that, for a planar detonation in a linear tube of circular cross section to initiate an unconfined spherical detonation, the tube must have a diameter d_c large enough to accomodate at least 13 transverse waves. Since the kinetic induction length is proportional to the transverse wave spacing, it should also be proportional to this critical tube diameter d_c. In Figure 4 we relate these quantities, together with experimental data (33-35). In these experiments, stoichiometric fuel-O_2 mixtures were diluted by successively larger amounts of N_2, until the ratio β of N_2/O_2 in the unreacted mixture was equal to 3.76, the value of β in normal air. In Figure 4 the experimental data are represented by symbols, with x indicating C_2H_2-oxidizer mixtures studied by Zeldovich et al. (35), plus indicating C_2H_4-oxidizer mixtures studied by Moen et al. (34), and all other symbols showing data of Matsui and Lee (33). Both conventional stoichiometric mixtures and "CO stoichiometric" mixtures are shown. The best agreement between computed and measured results is obtained when $d_c = 380\ \Delta$. Combined with the cell size proportionality from Figure 2 of a = 29 Δ, these computations predict

$$d_c = 13.1\ a \tag{1}$$

which is in excellent agreement with the observations that 13 cell diameters are required for the critical tube diameter. Because the overall agreement between computed predictions of d_c and experimental values shown in Figure 4 is good for all of the fuels considered (as well as for propane and hydrogen, not shown), the model can be used to predict values for fuel-oxidizer mixtures which have not been studied experimentally. It is also important to note that H_2-O_2 responds much differently to dilution by N_2 than do the hydrocarbon mixtures. Although H_2-O_2 has a value of Δ similar to that of C_2H_6-O_2, the computed variation of Δ with dilution for H_2 is much less than for the hydrocarbons, so that Δ for H_2-air is very similar to that of stoichiometric C_2H_2-air.

Critical tube diameters have also been measured for stoichiometric fuel-oxidizer mixtures at pressures different from atmospheric (33,36-38). Comparisons between experimental values for d_c and computed values of Δ at different initial

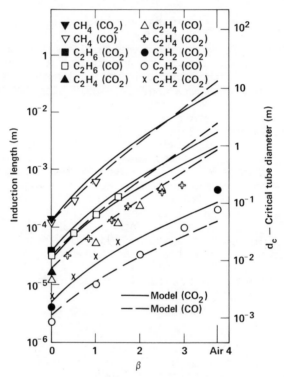

Figure 4.   Variation of computed induction length and critical tube diameter with degree of nitrogen dilution for stoichiometric fuel-oxidizer mixtures.   The symbols are described in the text.

pressures are shown in Figure 5. Again the general agreement is good when d_c = 380 Δ. Several other points are worth noting. First, for all of the hydrocarbon-O_2 mixtures,

$$P_0 = k_1 \Delta^{-\alpha} \tag{2}$$

is consistent with experimental observations (33). However, for H_2-O_2 there is a definite curvature in the computed results when they are plotted logarithmically as in Figure 5, although over the pressure range studied experimentally the apparently linear results are reproduced well by the kinetic model. All of the fuel-air computed results show curvature in Figure 5. Of particular interest is the computed curve for H_2-air which is multiple-valued for some values of d_c or Δ. Similar to explosion limit curves for H_2-O_2 (28), this behavior is due to changes with pressure in the dominant elementary reaction chains which occur in H_2 oxidation, particularly the competition between the reactions

$$H + O_2 \quad = O + OH$$
$$H + O_2 + M = HO_2 + M$$

No experimental data could be found to compare with this predicted behavior of H_2-air mixtures at different initial pressures; it would be interesting to see if the same value of d_c would be found at several different initial pressures, as the kinetic model predicts. Further, because of the curvature and multiple-valued behavior shown for some mixtures in Figure 5, Equation 2 should not be used for extrapolating results to pressures outside the range actually studied experimentally.

The model has also been used to predict the variation of induction length with initial gas temperature T_0 (35). At constant initial pressure, the induction length and critical tube diameter were found to increase slowly with T_0. The variation in initial density with T_0 is therefore more important than the slight increase in elementary reaction rates with T_0. These results suggest that the detonability of cold gas mixtures, such as those which can result from spills of liquefied natural gas (LNG) or other cryogenically stored fuels, will be slightly greater than the same mixtures at normal ambient temperatures.

Critical energy of direct initiation. The kinetics model has been used to predict the amount of energy necessary to initiate unconfined detonations, using the relation of Zeldovich et al. (35)

$$E_c \, \alpha \, \Delta^j \tag{3}$$

where j = 1, 2, or 3 for planar, cylindrical, or spherical

configurations, respectively. Comparisons between computed
values of Δj and critical high explosive initiator masses
(39-46) show good agreement (15,21-24). For example, in Figure
6 the critical mass of Tetryl high explosive required to initiate unconfined spherical detonation in C_2H_6-air (43) is
plotted as a function of equivalence ratio ϕ. Also shown is
the computed curve for Δ^3 for the same mixtures (21). The
shape of the curve and the value of ϕ corresponding to the
minimum value of E_c (and maximum detonability) are both reproduced very well by the numerical kinetic model. Similar agreement was found with the other fuel-air mixtures for which data
were available.

For initiation of spherical detonation by means of a linear
tube, Lee et al. (33,47) and Urtiew and Tarver (48) related the
critical initiation energy to the work which must be done to
produce a sufficiently strong source in the unconfined gas. The
resulting expression

$$E_c = k_0 \frac{P \; u}{D} d_c^3 \tag{4}$$

gives the critical energy in terms of the critical tube diameter
d_c and the pressure P, particle velocity u, and shock velocity
D of the CJ state. This can be combined with the relation
between d_c and Δ, giving

$$E_c = k_0 \frac{P \; u}{D} (380)^3 \Delta^3 \tag{5}$$

Computed values of E_c, based on the kinetic model (with k_0 =
0.1964) agree well (23,24) with values derived from experimental
measurements in fuel-O_2 mixtures (33,49).

For all of these correlations, perhaps the most significant
conclusion is that the computed curves agree simultaneously with
the experimental results for all of the fuels examined. The
ratios between the experimental and predicted values of E_c are
essentially the same in each case. This means that similar predictions can be made for other fuels for which experimental data
are lacking but for which a reliable kinetic mechanism exists.

Inhibition of detonation. Chemical additives can have a large
effect on experimentally observed ignition delay times. Small
amounts of NO_2, N_2O, H_2, or higher alkanes can significantly reduce the induction time in CH_4-O_2 and CH_4-air
mixtures (4,9,42,46,50-55). Conversely, halogenated species
have been shown (56,57) to increase the induction time in
fuel-O_2 mixtures. Since the induction time is a critical
factor in determining the detonability of fuel-oxidizer
mixtures, halogenated compounds and other species may have an
important role in reducing the detonability hazards of practical
fuels.

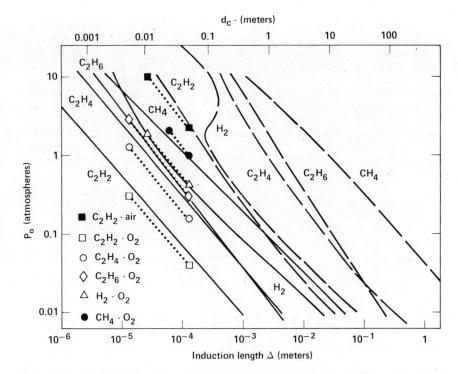

Figure 5. Variation of computed induction length and experimentally measured critical tube diameter with initial pressure. Solid curves are computed results for fuel-O_2 mixtures; dashed curves for fuel-air mixtures; and dotted lines with symbols represent experimental data from Matsui and Lee (33).

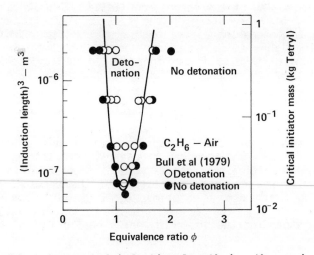

Figure 6. Cube of computed induction length in ethane-air mixtures, with data from Bull et al. (43).

Using the present model, the kinetic response of hydrocarbon-air mixtures to the addition of varying amounts of halogenated species has been examined (13). These inhibitors include the halogen acids HCl, HBr and $\overline{HI}$, as well as methyl, vinyl, and ethyl chlorides, bromides, and iodides. The common flame retardant CF_3Br was also used. As an example of these results, computed values of the induction length Δ are shown in Figure 7 for C_2H_4-air. From this figure it is clear that, relative to the case without inhibitors, all of the additives increased the induction length. This increase is smallest when Cl atoms are added, regardless of whether the additive is HCl, CH_3Cl, C_2H_3Cl, or C_2H_5Cl, and only the results with 1% HCl are shown in Figure 7. The iodides were most effective as inhibitors, and the bromides were nearly as effective as the iodides. The compound CF_3Br was slightly more effective as an inhibitor than CH_3Br, since the F atoms remove additional H atoms from the reacting mixtures, producing the relatively inert species HF. All of the inhibitors act as catalysts for the recombination of H atoms, lowering the size of the radical pool and reducing the rate of chain branching by means of the reaction

$$H + O_2 = O + OH$$

The rich limit for detonation in a 70 mm tube, measured by Borisov and Loban (27), is $\phi_R = 2.5$. At this point $\Delta = 1.04 \times 10^{-2}$ m. Based on the earlier discussion, the same value of Δ will correspond to the rich limit in the same tube for other mixtures as well. As inhibitors are added to the reactive fuel-air mixture, the value of equivalence ratio corresponding to $\Delta = 1.04 \times 10^{-2}$ m is gradually reduced. For 1% HI this gives a rich limit of $\phi_R \simeq 2.0$, showing a substantial narrowing of the detonation limits. Because the curves in Figure 7 are very steep on the fuel-lean side, most of the reduction in detonation limits occurs at the rich limit rather than at the lean limit. The increase in induction length with inhibitors also results in an increase in the critical energy for initiation of detonation, which can be seen easily from Equation 5.

Conclusion

Detonations are extremely complex phenomena and involve many competing physical and chemical processes. Complete theoretical models of the initiation, stability, and structure of detonation waves require an accurate description of the chemical kinetics of the induction phase. The primary goal of the present work is to demonstrate that kinetic mechanisms are now available which are able to predict the induction delay period for a variety of practical fuels.

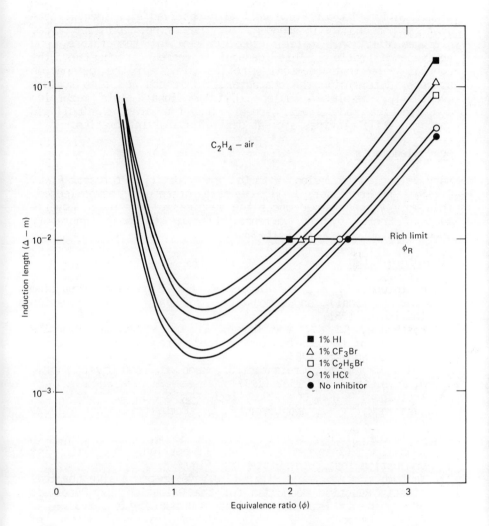

Figure 7. Computed induction lengths for ethylene-air mixtures showing the effects of addition of selected inhibitors. Also shown is the predicted rich limit for the propagation of detonation in a linear tube, based on the data without inhibitors present of Borisov and Loban (27).

It was observed that computed induction delay times and induction lengths correlate well with observed experimental detonation phenomena, independent of fuel type, oxidizer type, nitrogen dilution, initial pressure, initial temperature, and equivalence ratio. This general agreement emphasizes the central role that chemical kinetics plays in the detonation process, determining the characteristic length and time scales. Eventually, complete coupled multidimensional fluid mechanics and kinetics models will appear, but the present simplified approach still provides a great deal of useful information.

Acknowledgments

Many valuable discussions with Dr. P. A. Urtiew, Professor F. L. Dryer, and Professor J. H. Lee are gratefully acknowledged. This work was performed under the auspices of the U. S. Department of Energy by the Lawrence Livermore National Laboratory under contract No. W-7405-ENG-48.

Literature Cited

1. Westbrook, C. K., and Dryer, F. L., Eighteenth Symposium (International) on Combustion, p. 749, The Combustion Institute, 1981.
2. Westbrook, C. K., and Dryer, F. L., "Chemical Kinetics Modeling of Hydrocarbon Combustion", Prog. Energy and Comb. Sci., to appear, 1983.
3. Westbrook, C. K., Creighton, J., Lund, C., and Dryer, F. L., J. Phys. Chem. $\underline{81}$, 2542 (1977).
4. Westbrook, C. K., Comb. Sci. Tech. $\underline{20}$, 5 (1979).
5. Westbrook, C. K., and Dryer, F. L., Comb. Sci. Tech. $\underline{20}$, 125 (1979).
6. Westbrook, C. K., Dryer, F. L., and Schug, K. P., Nineteenth Symposium (International) on Combustion, p. 153, The Combustion Institute, Pittsburgh, 1983.
7. Westbrook, C. K., and Pitz, W. J., "A Comprehensive Chemical Kinetic Reaction Mechanism for the Oxidation and Pyrolysis of Propane and Propene, submitted for publication, 1983.
8. JANAF Thermochemical Tables, U. S. National Bureau of Standards NSRDS-NBS 37 and Supplements. D. R. Stull and H. Prophet, eds., 1971.
9. Westbrook, C. K., and Haselman, L. C., Prog. in Astr. and Aero. $\underline{75}$, 193 (1981).
10. Jachimowski, C. J., Comb. Flame $\underline{29}$, 55 (1977).
11. Burcat, A., Lifshitz, A., Scheller, K., and Skinner, G. B., Thirteenth Symposium (International) on Combustion, p. 745, The Combustion Institute, Pittsburgh, 1971.
12. Westbrook, C. K., Comb. Sci. and Tech. $\underline{23}$, 191 (1980).
13. Westbrook, C. K., Nineteenth Symposium (International) on Combustion, p. 127, The Combustion Institute, Pittsburgh, 1983.

14. Westbrook, C. K., "Numerical Modeling of Flame Inhibition by CF₃Br", Comb. Sci. and Technol., in press (1983).
15. Westbrook, C. K., Comb. Sci. and Tech. 29, 65 (1982).
16. Bull, D. C., Elsworth, J. E., Schuff, P. J., and Metcalfe, E., Comb. Flame 45, 7 (1982).
17. Strehlow R. A. and Engel, D. C., AIAA J. 7, 492 (1969).
18. Strehlow, R. A. and Rubins, P. M., AIAA J. 7, 1335 (1969).
19. Manzhalei, V. I., Mitrofanov, V. V., and Subbotin, V. A., Fiz. Goren. Vzr., 10, 102 (1974).
20. Vassiliev, A. A., Fiz. Goren. Vzr. 18, 132 (1982).
21. Westbrook, C. K., Comb. Flame 46, 191 (1982).
22. Westbrook, C. K., and Urtiew, P.A., Fiz. Goren. Vzr., in press, 1983.
23. Westbrook, C. K., Pitz, W. J., and Urtiew, P. A., "Chemical Kinetics of Propane Oxidation in Gaseous Detonations", submitted for publication, 1983.
24. Westbrook, C. K., and Urtiew, P. A., Nineteenth Symposium (International) on Combustion, p. 615, The Combustion Institute, Pittsburgh, 1983.
25. Pawell, D., Vasatko, M., and Wagner, H. Gg., "The Influence of Initial Temperature on the Limits of Detonability", AFOSR 69-1095TR AD-692900, 1967.
26. Kogarko, S. M., Soviet Physics-Tech. Phys. 3 (28), 1904 (1958).
27. Borisov, A. A., and Loban, S. A., Fiz. Goren. Vzr. 13, 729 (1977).
28. Lewis, B. and von Elbe, G., Combustion, Flames and Explosions of Gases, Academic Press, New York, 1961.
29. Michels, H. J., Munday, G., and Ubbelohde, A. R., Proc. Roy. Soc. London A 319, 461 (1970).
30. Mitrofanov, V. V., and Soloukhin, R. I., Sov. Phys. Dokl. 9, 1055 (1964).
31. Edwards, D. H., Thomas, G. O., and Nettleton, M. A., J. Fluid Mech. 95, 79 (1979).
32. Lee, J. H., Knystautas, R., and Guirao, C. M., Proceedings of the International Specialist Meeting on Fuel-Air Explosion, Montreal, Canada, p. 157, University of Waterloo Press, 1982.
33. Matsui, H., and Lee, J. H., Seventeenth Symposium (International) on Combustion, p. 1269, The Combustion Institute, Pittsburgh, 1979.
34. Moen, I. O., Donato, M., Knystautas, R., and Lee, J. H., Eighteenth Symposium (International) on Combustion, p. 1615, The Combustion Institute, Pittsburgh, 1981.
35. Zeldovich, Ya. B., Kogarko, S. M., and Semenov, N. N., Sov, Phys. Tech. Phys. 1, 1689 (1956).
36. Vassiliev, A. A., and Grigoriev, V. V., Fiz. Goren. Vzr. 16, 177 (1980).
37. Vassiliev, A. A., Fiz. Goren. Vzyrva 18, 98 (1982).
38. Knystautas, R., Lee, J. H., and Guirao, C. M., Comb. Flame 48, 63 (1982).

39. Atkinson, R., Bull, D. C., and Shuff, P. J., Comb. Flame 39, 287 (1980).
40. Bull, D. C., Trans. I. Chem. Eng. 57, 219 (1979).
41. Bull, D. C., Elsworth, J. E., and Hooper, G., Acta Astr. 5, 997 (1978).
42. Bull, D. C., Elsworth, J. E., and Hooper, G., Comb. Flame 34, 327 (1979).
43. Bull, D. C., Elsworth, J. E., and Hooper, G., Comb. Flame 35, 27 (1979).
44. Carlson, G. A., Comb. Flame 21, 383 (1973).
45. Nicholls, J. A., Sichel, M., Gabrijel, Z., Oza, R. D. and Vander Molen, R., Seventeenth Symposium (International) on Combustion, p. 1223, The Combustion Institute, Pittsburgh, 1979.
46. Vander Molen, R., and Nicholls, J. A., Comb. Sci. Tech. 21, 75 (1979).
47. Lee, J. H. and Matsui, H., Comb. Flame 28, 61 (1977).
48. Urtiew, P. A., and Tarver, C. M., Prog. Astr. Aero. 75, 370 (1981).
49. Vassiliev, A. A., Nikolaev, Yu. A., and Ulianitski, V. Yu., Fiz. Goren. Vzyrva 15, 94 (1979).
50. Burcat, A., Comb. Flame 28, 319 (1977).
51. Slack, M. W., and Grillo, A. R., Comb. Flame 40, 155 (1981).
52. Dorko, E. A., Bass, D. M., Crossley, R. W., and Scheller, K., Comb. Flame 24, 173 (1975).
53. Dabora, E. K., Comb. Flame 24, 181 (1975).
54. Crossley, R. W., Dorko, E. A., Scheller, K., and Skinner, G. B., Comb. Flame 19, 373 (1972).
55. Eubank, C. S., Rabinowitz, M. J., Gardiner, W. C., Jr., and Zellner, R. E., Eighteenth Symposium (International) on Combustion, p. 1767, The Combustion Institute, Pittsburgh, 1981.
56. Skinner, G. B., and Ringrose, G. H., J. Chem. Phys. 43, 4129 (1965).
57. Skinner, G. B., Halogenated Fire Suppressants, ACS Symposium Series 16, R. G. Gann (ed.). American Chemical Society, 295 (1975).

RECEIVED October 28, 1983

Review of Plasma Jet Ignition

R. M. CLEMENTS

Department of Physics, University of Victoria, Victoria, B.C., V8W 2Y2, Canada

During the last decade or so there has been considerable interest in lean-burn, spark ignited, internal
combustion engines. Motivating reasons behind this
are the reduction of pollutants and increased fuel
economy. However such lean mixtures are difficult
to ignite and have longer burn times by comparison
to stoichiometric mixtures. One way of alleviating
these problems is with plasma jet ignition. The
mechanical construction of typical igniters as well
as associated electrical circuitry required to power
igniters are discussed. The igniter produces a puff
of gas which is the ignition kernel. How this puff
behaves is examined when it is either in the
presence or absence of a combustible mixture. This
is done in the light of the relative roles of fluid
mechanical turbulence and chemical effects. Finally,
the practical aspects of using these igniters in
internal combustion engines as well as associated
problems are discussed.

Lean burn, spark ignited, internal combustion (ic) engines
especially as used in motor vehicles have received considerable
attention during the past few years. The reasons for this are
well known: by comparison with stoichiometric or rich mixtures,
lean mixtures produce a lower pollutant level and improved fuel
economy. There are problems associated with lean burn,
specifically the mixture is slow burning and hard to ignite.
There are numerous proposals for the alleviation of these
problems, including charge stratification, high levels of
turbulence or swirl induced by special valve or piston design, as
well as improved electrical ignition systems. These are reviewed
by Dale and Oppenheim (1). One such form of enhanced ignition
system is plasma jet ignition (PJI) and this is the subject of
the present review. Because a companion paper in this symposium

0097-6156/84/0249-0193$06.00/0
© 1984 American Chemical Society

series (Sloane and Ratcliffe (2)) concentrates on the chemistry of
ignition the present paper will emphasize the role of fluid
mechanics with the chemical role somewhat subordinated. Finally
because of the large number of research papers in the area of PJI
the present review will be aimed toward general concepts and no
attempt will be made to review every paper published in this
field.

Basic PJI System

Figure 1 shows the simplest type of plasma jet igniter of the
design used by Topham et al (3) and Asik et al (4). While there
is a wide variation in the different designs, the design indicated
in Figure 1 is typical. However there are two interesting design
variations. The first entails placing a nozzle at the jet orifice
(see for example Oppenheim et al (5)). The second is the
incorporation of a gas inlet into the small blind chamber of the
igniter, as proposed by Weinberg et al (6). The latter means that
the chamber of the plasma igniter need not contain the same gas
mixture as that outside the plasma jet cavity.

Although somewhat different in design details, all of these
igniters work in the same manner. An arc strikes between the
center electrode and the end plate. This arc heats the gas in the
cavity and the overpressure expels a thin jet of highly luminous
plasma. It should be noted that there is essentially no
interaction between the current in the arc and its own self-
produced magnetic field. This was not the situation in the
igniters investigated by Bradley and Critchley (7) or Harrison and
Weinberg (8).

Although the circuits to power the igniter used by the
different investigators (1, 3-5) differ in detail they are
basically similar. The circuit consists essentially of two
stages. The first stage produces a high voltage (10s of kV) low
energy (100 mJ typically) pulse which causes the initial breakdown
of the gap in the plasma igniter. A low voltage (say < 1 kV)
maintains the electrical discharge dissipating energies in the
order of a few joules with maximum currents between 100A and
1000A. Typically the time to current maximum is in the order of a
few 10s of μs. Streak photographs (3, 5) have shown that the
luminous plasma jet is a few centimeters long and after current
maximum the jet quickly becomes non-luminous. The velocity of the
front of this luminous jet ranges from about half the speed of
sound to slightly over the speed of sound for these conditions.
These streaks often also show considerable structure behind the
luminous front. Some of this structure can be interpreted as
gaseous material ejected from the igniter at velocities
considerably higher than that of the luminous front.

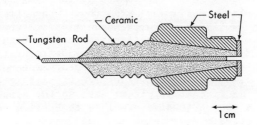

Figure 1.   A cross section of a typical simple plasma jet igniter.
(Reproduced with permission from Ref. 3.   Copyright 1975, The
Combustion Institute.)

The Gaseous Puff Produced by PJI and Ignition

For typical combustion reactions induction times are in the
order of milliseconds. However for many of the experiments
discussed above, the luminous jet produced by the igniter dies
away in times an order of magnitude shorter than this quantity.
In order to investigate this, time resolved Schlieren photography
is employed. This was first done by Oppenheim et al (5) and
subsequently by other workers. Figure 2 (A) shows photographs of
the television screen of a time resolved Schlieren system for a
plasma jet igniter into air at four atmospheres pressure. The
puff of warmish gas which exists long after all plasma jet
luminescence has disappeared is noteworthy and appears very much
like an atmospheric thermal or a turbulent puff as described by
Richards (9). In fact it is mathematically similar to the
turbulent puff described by Richards and the relationships which
he derives for his large scale turbulent puffs apply to the puff
generated by the plasma jet. Thus even though the time and length
scale between the two types of puffs may differ up to orders of
magnitude the same mathematical analysis describes both (10).
        Clearly, as discussed in some detail by Orrin et al (11),
fluid mechanical effects are not the whole story when one is using
the plasma jet as a source of ignition. Figure 2(B) show a series
of photographs of ignition by a plasma jet and for comparison
(Figure 2(C)) by a conventional spark in a combustion bomb
containing a methane-air mixture at four atmospheres pressure.
For these pictures the normalized air-fuel ratio $\lambda = 1.3$ and for
all cases the combustion chamber dimensions are 3 cm × 4 cm ×
9 cm. The striking similarity of the plasma jet with and without
combustion is very evident from the photographs. This would tend
to imply that at least for this mixture strength turbulence is
governing the combustion process, or at least the early stages
thereof. Also the dramatic difference between a conventional
spark ignition and the plasma jet ignition is clearly shown and it
is evident that the combustion for spark ignition is considerably
slower than that for the plasma jet. This is again consistent
with the high levels of turbulence generated by the plasma jet and
their affect on the combustion process.
        Because the Schlieren photographs shown in Figure 2(B) and
(C) are at best quasi-quantitative the more quantitative
measurement of pressure inside the combustion chamber versus time
has been made.   Figure 3 shows such a graph for the same
conditions as shown in Figure 2(B) and (C). The difference
between conventional spark ignition and PJI are very evident. To
facilitate comparisons between the two systems for different
values of air-fuel ratio the following parameters are defined.
Delay time is taken as the time from the start of the electrical
pulse to the igniter to the time when the pressure reaches 10% of
its maximum value and burn rate is taken as the mean slope between
10% maximum pressure and 90% maximum pressure. Figure 4 gives

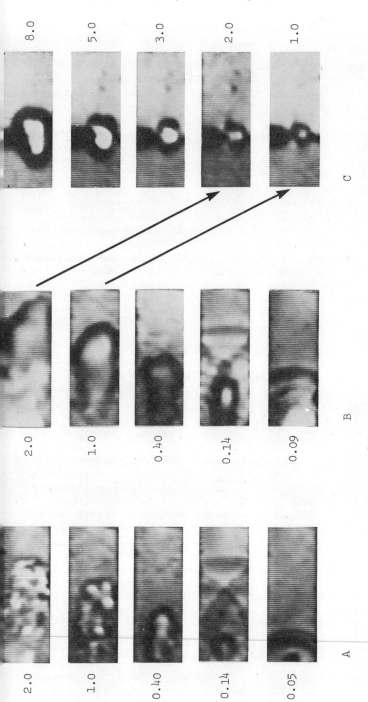

Figure 2. Time-resolved Schlieren photographs of a constant volume combustion chamber at 4 atm. pressure. The vertical dimension of each frame is 0.3 cm. A, plasma jet igniter into air; B, plasma jet ignition of a methane–air mixture with λ = 1.3; and C, conventional spark ignition of a methane–air mixture with λ = 1.3. Numbers are time in ms.

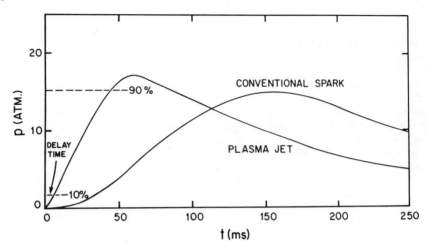

Figure 3. A pressure (p) time (t) history of combustion in a constant volume chamber for a methane-air mixture with λ = 1.3. (Reproduced with permission from Ref. 20. Copyright 1983, Combustion Science and Technology.)

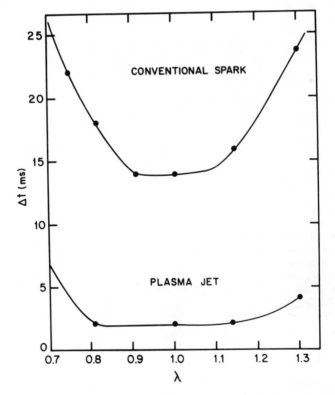

Figure 4. Induction delay vs. λ for a methane-air mixture in a constant volume chamber. (Reproduced with permission from Ref. 20. Copyright 1983, Combustion Science and Technology.)

delay time versus λ while Figure 5 gives the burn rate versus λ. The difference between the conventional spark and PJI are very evident.

All the preceding data are not for ultra lean mixtures (i.e. mixtures near or beyond the normal lean limit of flamability). Also the gas inside the chamber of the plasma jet igniter is the same as the mixture in the main combustion chamber. This data is compatible with the premise that the plasma jet causes a high degree of turbulence which increases the flame front area and this accounts for the reduction in delay time and increase in burn rate.

In an initial brief report by Weinberg et al (6) and its subsequent expansion by Orrin et al (11) the effect of different gases and liquids in the plasma jet cavity was investigated. They worked at both constant volume (6) and constant pressure (11). An igniter of the basic design which is attributed to them in the preceding section was used to ignite lean (λ in the order of 2) methane-air mixtures. The igniter cavity was filled with different gases (for example hydrogen, methane, argon, etc.) and different liquids (for example alcohols, aldehydes, benzene, water etc). The size of the combusting region was measured using Schlieren techniques and the concentration of numerous interesting species, for example H, OH, O, etc. were also measured. It is clear that the high energies developed in a plasma jet will generate numerous radicals which are of interest in the ignition of the combustion reaction. Their goal was to find out which were most important and also to separate the fluid mechanical effects which have been discussed previously in the present paper, from chemical effects. Briefly stated, the conclusions were that for these very lean mixtures atomic hydrogen is of utmost importance and that roughly speaking (at least up to a given minimum hydrogen concentration in the species contained in the plasma jet cavity) the more hydrogen the species contained the better the ignition properties were. For these very lean mixtures they found that argon was 'an exceptionally poor ignition source'. Inasmuch as Zhang et al (12) found that for mixtures near stoichiometric argon was a very good source of ignition, one can surmise, based on all the data presented here, that for mixtures near stoichiometric fluid mechanical effects dominate and chemical effects are subordinated probably because there are the needed concentration of radicals present no matter what one uses as a gas in the plasma jet cavity. However for extremely lean (6, 11) mixtures the chemical effects predominate and in fact a high degree of turbulence may well be counterproductive to effective ignition. As a final comment: it is not inconceivable that other radical species, as yet not investigated, could play a major role in the ignition process.

PJI in Engines

There have been a number of investigations (4, 13-16) on the

effects of PJI on the performance of the internal combustion
engine. The investigations have considered not only the usual
liquid fuel gasoline, but also the gaseous fuels methane and
propane. Most of the investigations, but not all (4), have been
done on single cylinder laboratory test engines the most popular
being a CFR engine. Results from these investigations are
essentially mutually consistent and thus rather than discuss each
of the experiments individually a typical one will be discussed in
a little detail. Also of the fuels discussed methane is the
simplest and shows the strongest effects when using PJI. Hence,
for simplicity the experiment by Pitt and Clements (15) will be
discussed.

The engine used in this experiment (15) was a single cylinder
L-head test engine connected to an electric dynamometer. The
standard comparison ignition system was a capacitive discharge
system (CDI) storing about 30 mJ and the PJI system storing 1.2 J.
A low compression ratio of 4:1 was used in order to avoid any
problems of electrical misfire with the plasma jet. The engine
displacement was about 500 cm^3 and all tests were taken at 2000
rpm and a constant methane flow of 0.43 l/s. The engine cylinder
pressure histories were measured by a piezo-electric pressure
transducer. A summary of the results is as follows. Typically
the CDI system required 20° more timing advance than did the PJI
system for optimal performance of both systems. However when
conditions were not optimized the PJI system showed considerably
less cycle-to-cycle variation of the pressure history. Also PJI
extended the lean misfire limit from $\lambda = 1.2$ to slightly beyond
$\lambda = 1.3$.

For the conditions discussed in the preceding paragraph, even
for optimal conditions, there was some cycle-to-cycle variation in
the pressure history. Thus to obtain meaningful results the
average of a large number (say 100 or more) of pressure traces
must be taken. A great deal of this variation is because the air-
fuel ratio varies from cycle to cycle. This results from the fact
that the burn and scavenging action of the previous cycle affects
the air-fuel ratio for the cycle in question. In order to
alleviate this problem the engine was motored at the desired speed
(2000 rpm) and the air-fuel ratio stabilized at the desired value.
The ignition was fired only once and the cylinder pressure history
recorded. Figure 6 is obtained from the average of ten such
"shots" and as usual gives log pV^γ as a function of crank angle.
Here p is cylinder pressure, V is cylinder volume and γ is the
ratio of specific heats. One sees from this figure that there is
essentially no difference between the two curves except that an
additional 20° of advance is required for the CDI system. In this
figure optimal timing is chosen for both ignition systems and the
mixture is stoichiometric. However, for lean mixtures (say $\lambda >$
1.2) there is a difference between the standard ignition system
and the PJI system during the combustion phase; the PJI system
causing a faster burn.

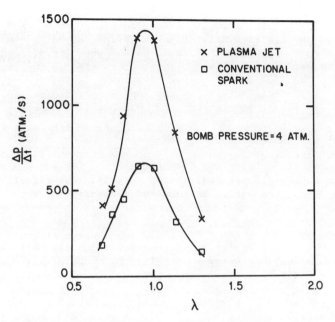

Figure 5. Burn rate ($\Delta p/\Delta t$) vs. λ for a methane-air mixture in a constant volume chamber. (Reproduced with permission from Ref. 20. Copyright 1983, Combustion Science and Technology.)

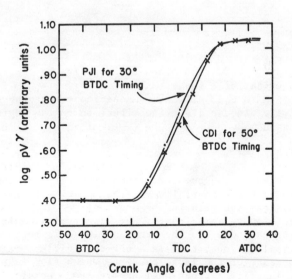

Figure 6. Log pV^{γ} vs. timing angle for combustion of a methane-air mixture ($\lambda = 1.0$) in a single-cylinder test engine. (Reproduced with permission from Ref. 15. Copyright 1983, Combustion Science and Technology.)

In a manner similar to that done for constant volume combustion the ignition delay time is defined as being the time between when the ignition system fires and when log pV^γ reaches 10% of its maximum value, and also the burn time is the time between 10% and 90% on the log pV^γ diagram. Table I shows these times for both ignition systems and for two air-fuel ratios as a function of timing. Again one sees an appreciable decrease in the

Table I. Ignition Delay and Combustion Times for log pV^γ analysis (Reproduced with permission from Ref. 15. Copyright 1983, "Combustion Science and Technology".)

Spark Time (°BTDC)	Air-Fuel Ratio	PJI		CDI	
		Ignition Delay (ms)	Combustion Time (ms)	Ignition Delay (ms)	Combustion Time (ms)
20	1.0	1.3	1.9	2.6	2.0
30	1.0	1.4	2.1	2.5	2.0
40	1.0	1.4	2.1	2.8	2.0
50	1.0	1.8	2.0	3.2	2.1
40	1.2	2.0	2.1	3.0	2.5
50	1.2	2.3	2.2	3.3	2.8
60	1.2	2.4	2.3	3.6	2.8
70	1.2	2.3	2.3	4.0	2.6
80	1.2	-	-	4.4	2.0

ignition delay with PJI but minimal change in the combustion time for the two values of λ shown.

Thus, in a more quantitative manner, one sees the same results as were discussed for the situation of the engine running in a continuous mode.

Problems with PJI

There are basically two problems associated with PJI and these are inter-related. For essentially all the experiments reviewed in the present paper energies in the order of 1 J are dissipated each time the ignition system fires. Thus in a typical multi-cylinder engine electrical power in the order of 100 W is required for, say automotive applications; this is a non-negligible amount of power. Secondly, because of the large currents which flow in the plasma jet igniter there is a great deal of electrode erosion. That this is so is no surprise in that it has long been known (17) that for every coulomb of charge passed between two electrodes a certain

amount of electrode material is eroded. These two problems have
been considered by Smy et al. (18) and they conclude that igniter
lifetimes in the order of 10 or so hours are to be expected.

Obviously the simplest solution to the above two problems is
to reduce the amount of energy dissipated in the igniter. Thus
at present there are investigations aimed at determining the
minimum amount of energy which needs to be delivered to the
igniter in order to generate a properly formed igniting puff and
also the temporal manner in which this energy should be
deposited. Oppenheim (19) has used a small watch jewel as the
orifice in the plasma jet and thus reduced the erosion of the end
plate of the plasma jet. This concept coupled with electrode
materials which are difficult to erode may well mitigate the
lifetime problem as related to the plasma jet. Finally, it
appears possible to mechanically generate a puff which is in the
fluid mechanical sense similar to that generated by a plasma jet.
Pitt et al (20) have used a fast acting mechanical valve to
generate the highly turbulent puff. This puff then passes between
two low energy conventional spark electrodes. The result is very
similar to PJI and such a system requires very little electrical
energy.

Conclusions

For PJI ignition of non turbulent mixtures it appears that for
values of $\lambda \approx 1$ fluid mechanical effects dominate the development
of the ignition stage while for values of λ near or beyond the
conventionally defined lean flammability limit it is chemical
effects which dominate. For highly turbulent mixtures, i.e.
internal combustion engines, when the mixture strength is near
stoichiometric and for optimal timing of each ignition system
there appears to be minimal difference between them. However for
lean mixtures PJI extends the lean limit appreciably. For all
situations PJI is much less sensitive to timing than is the
conventional ignition system. This may well be useful in dual
fuel vehicles, for example gasoline and methane fueled vehicles.
In this circumstance it is normally necessary to advance the
ignition some 20° when changing from gasoline to methane as a
fuel. With PJI an optimal timing could be chosen which would
cause good combustion for either fuel and the timing advance left
unchanged. Finally from a practical point of view the problems
related to high energy consumption and high electrode wear must be
overcome before PJI can become a commercially viable ignition
source.

Acknowledgments

This work was supported by National Sciences and Engineering
Research Council of Canada as well as the British Columbia Science
Council.

Literature Cited

1. Dale, J.D.; Oppenheim, A.K.   SAE paper No. 810146, 1981.
2. Sloane, R.M.; Ratcliffe, J.W.   "Experimental and Computational Studies of the Chemistry of Ignition Process", this symposium.
3. Topham, D.R.; Smy, P.R.; Clements, R.M.   Combust. Flame, 1975, 25, 87.
4. Asik, J.R.;, Piatkowski, P.; Foucher, M.J.; Rado, W.G.   SAE paper No. 770355, 1977.
5. Oppenheim, A.K.; Teichman, K.; Hom, K; Stewart, H.E. SAE paper No. 780637, 1978.
6. Weinberg, F.J.; Hom, K.; Oppenheim, A.K.; Teichman, K. Nature 1978, 272, 341.
7. Bradley, D.; Critchley, I.L. Combust. Flame 1974, 22, 143.
8. Harrison, A.J.; Weinberg, F.J. Combust. Flame 1974, 22, 263.
9. Richards, J.M. J. Fluid Mech. 1965, 21, 97.
10. Topham, D.R.; Zhang, J.X.; Clements, R.M.; Smy, P.R. J. Phys. D: Appl. Phys. 1982, 15, L65.
11. Orrin, J.E.; Vince, I.M.; Weinberg, F.J. in "Eighteenth Symposium (International) on Combustion", The Combustion Institute: Pittsburgh, 1981, p. 1755.
12. Zhang, J.X.; Clements, R.M.; Smy, P.R. Combust. Flame 1983, 50, 99.
13. Dale, J.D.; Smy, P.R.; Clements, R.M. Combust. Flame 1978, 31, 173.
14. Tozzo, L.; Dabora, E.K. "Nineteenth Symposium (International) on Combustion", The Combustion Institute: Pittsburgh, 1982, p. 1467.
15. Pitt, P.L.; Clements, R.M. Combustion Sci. and Tech. 1983, 30, 327.
16. Edwards, C.F.; Dale, J.D.; Oppenheim, A.K. SAE paper No. 830479, 1983.
17. Cobine, J.D. "Gaseous Conductors", Dover Publications, 1958; p. 301.
18. Smy, P.R.; Clements, R.M.; Dale, J.D.; Simeoni, D.; Topham, D.R. J. Phys. D: Appl. Phys. 1983, 16, 783.
19. Oppenheim, A.K., personal communication.
20. Pitt, P.L.; Ridley, J.D.; Clements, R.M. Combustion Sci. and Tech., accepted for publication.

RECEIVED October 26, 1983

Chemistry of Spark Ignition
An Experimental and Computational Study

THOMPSON M. SLOANE and JOHN W. RATCLIFFE

Physical Chemistry Department, General Motors Research Laboratories, Warren, MI 48090

The chemistry of ignition has been investigated by
time-resolved molecular beam mass spectrometry and
schlieren photography. A simultaneous characterization of the physical and chemical properties of
flame ignition has yielded the first detailed chemical information about the process. Time-of-flight
detection has allowed a correspondence to be made
between the oxygen signal detected as a function
of time after the spark and the oxygen concentration at the sampling cone tip. Flame propagation
appears to begin about 0.2 cm from the spark electrodes rather than between the electrodes as has
been conventionally assumed. A one-dimensional
combustion model shows that heat loss to the electrodes can cause the flame to begin at a distance
from the electrodes rather than in the electrode

We have previously demonstrated (1) our ability to make time-
resolved mass spectrometer sampling measurements of a flame
propagating through a combustion bomb. Our interest in making
these kinds of measurements is to study the chemistry of ignition by various ignition methods and to determine the value of
these different methods for combustion modification. This report
presents our first results on the chemistry of spark ignition.

There are at least two reasons why a detailed study of
spark ignition is worthwhile. First, very little is known about
the chemistry of spark ignition. The identity and quantity of
radicals produced in different mixtures compared to the amount
of energy deposited as heat could be an important factor in the
ignitability of these mixtures. An improved method of ignition
could decrease the hydrocarbon emissions from a direct-injection

0097-6156/84/0249-0205$06.00/0

stratified charge engine and could also remedy problems due to
wet or fouled spark plugs in homogeneous spark ignited engines.
Second, knowledge of the chemistry of spark ignition serves as a
baseline of comparison with other types of ignition such as
plasma jet and photochemical ignition. The differences observed
among these different types of ignition is related to the differences in the chemistry and fluid mechanics associated with
them. The present study therefore forms an important step in
our investigation into the mechanism of combustion initiation
with the intent to modify the combustion process in a favorable
way through the ignition process.

There is very little known about the chemistry of ignition
by electric spark, or by any other means for that matter. A
number of physical characterizations have given rise to a generally accepted overall picture of how ignition proceeds. One such
description of what is known about spark ignition is given in
Lewis and von Elbe (2). Within the small volume of heated gas
produced by the spark there is undoubtedly some ionization and
dissociation in addition to thermal heating. It is not known
what distribution of energy in these different forms provides an
optimum mix for ignition. It is apparent, though, that a minimum
amount of energy is necessary to produce a flame kernel of a
certain critical size. If this critical size is not attained,
the flame does not propagate.

The results obtained here give a qualitative picture of the
evolution of chemical components for an electric spark in three
mixtures, two of which ignite and one of which does not ignite.
We will show the time and spatial evolution of the concentration
of an important radical as well as reactants. We also show
simultaneous flash Schlieren pictures of the ignition to charac-
terize the physical aspects of the ignition process and of the
sampling method. Our first results of time-of-flight detection
which yield greatly enhanced time resolution will be presented
for oxygen. A comparison of these experiments with the unchopped
signal enhances the value of the data presented.

Experimental

The apparatus used in the molecular beam sampling experiments has
been previously described in detail (1). A schematic diagram is
shown in Figure 1. A quartz cone having an included angle of 60°
at the tip mounted on the wall of the combustion bomb samples the
region of interest by serving as the nozzle of a molecular beam
sampling system. A multichannel analyzer records mass spectrometer signals as a function of time after the spark. The spark is
2 μs in duration and is produced by a circuit which is supplied
with 200 mJ of electrical energy. The location of the 0.1 cm
spark gap can be changed so that different regions of the developing flame can be sampled. Filling and emptying of the bomb,

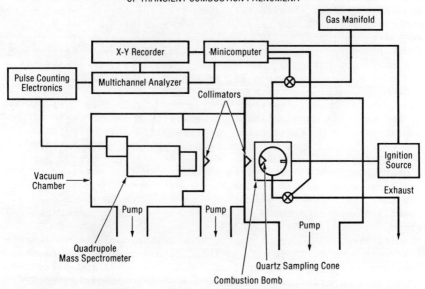

Figure 1.   Schematic diagram of the apparatus.   (Reproduced with permission from Ref. 1.   Copyright 1983, Combust. Sci. Technol.)

spark timing, and data collection are performed under the control of a microcomputer so that signal can be obtained from
repetitive ignitions. The bomb was filled to an initial pressure of 32 kPa with mixtures of CH_4-O_2-Ar having different
stoichiometries. The ratio of oxygen to argon was identical to
the O_2:N_2 ratio in air for all but one experiment.

In order to increase the time resolution in these experiments, a time-of-flight chopper was placed 6.9 cm downstream from
the cone tip, between the bomb and the skimmer. The chopper
blade has four equally spaced slots and is driven by a 400 Hz
synchronous motor. The resulting shutter function gives an open
time of 50 μs followed by a closed time of 575 μs. Taking into
account the flight time distribution from the sampling cone tip
to the chopper, signals can then be obtained with a time resolution of about 100 μs at 625 μs intervals for a given time relationship between the chopper position and the spark. This type
of measurement has been made for argon and oxygen in a stoichiometric mixture with the tip of the cone located 0.5 cm and 0.2
cm from the spark electrodes.

We considered it likely that placing the chopper between the
nozzle and the skimmer might interfere with the supersonic expansion downstream of the nozzle. We were pleasantly surprised to
find that this was not the case because we obtained velocity
distributions with the chopper located between the nozzle and the
skimmer which were identical to the distributions obtained when
the chopper was located in the more conventional position downstream of the skimmer. This is undoubtedly due to the non-
ideality of the flow through the cone which will be described
later.

Schlieren photographs of the spark kernel and the developing
flame were obtained using a conventional arrangement. A flashlamp-pumped dye laser was used to furnish a 1 μs light pulse
($\lambda \approx 600$ nm) to illuminate the interior of the combustion bomb.
By delaying the laser pulse for a variable time after the spark,
the progress of the ignition kernel and the flame could be
followed at very short delay times for sequential ignitions. A
series of pictures which shows the flame approaching and then
passing the cone tip is shown in Figure 2. There is no noticeable distortion of the flame as it approaches the tip because the
flame shape in the absence of the probe is virtually identical to
the flame shape with the probe present. This is consistent with
the finding of Yoon and Knuth ([3]) that at comparable pressure in
a steady-state flame sampling experiment the cone has no appreciable effect on the flame. The length of the orifice is short
enough (~0.015 cm or twice the diameter) that there should be
little radical recombination in the orifice ([3]).

The velocity distribution of argon sampled with the quartz
cone at 32 kPa and 300 K, the initial pressure and temperature,
corresponds to a terminal Mach number of about 2.8 in the molecular beam, which is somewhat smaller than the expected Mach

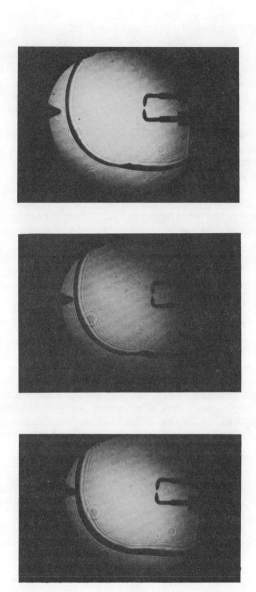

Figure 2.   Schlieren photographs of a stoichiometric flame passing the sampling cone.

number of 16.6. This is very likely due to the presence of a
boundary layer along the inside walls of the sampling cone which
prevents much of the cooling which accompanies the isentropic
expansion of the sampled gas. When the cone is replaced by a
simple free jet source having the same orifice size, the Mach
number characterizing the velocity distribution obtained is about
15. Future experiments will be performed with cones having a
larger included angle to attempt to minimize this problem.
According to the work of Yoon and Knuth, we should be able to use
a sampling cone with an angle of 110° without unduly disturbing
the flame. Preliminary experiments with such a cone yield
velocity distributions which are nearly identical to that obtained with the free jet source. This has confirmed our sus-
picion that the smaller angle cone used in these experiments
causes a thermalization of the beam in a thick boundary layer
which fills the region of the cone just downstream of the orifice.
This boundary layer may be present with the 110° cone, but it
apparently does not interfere with the core of the jet because of
the larger angle. We expect to achieve higher beam intensities,
narrower velocity distributions, and less velocity slip, i.e.,
less difference in the velocity of different components, with
this cone. Perhaps the most important advantage to be gained is
a narrow velocity distribution because this has a direct effect
on the time resolution in these experiments.

     In view of the fact that the sampling conditions are some-
what less than ideal, time-dependent signals obtained for radicals must be viewed with some caution. The time dependence of
the major components, however, should be unaffected by the low
Mach number flow through the sampling cone. Subsequent experiments with the 110° cone should provide much more definitive information on radical relative mole fractions, temperature, and
perhaps also absolute mole fractions of a number of components.

Results

Schlieren Photography of the Ignition Kernel. Schlieren photographs taken at very short times after the spark (3-44 μs) show
that although the cone does cause a disturbance of the shock wave
produced by the spark when it is located 0.2-0.5 cm from the
electrodes, there is very little visible effect on the development of the spark kernel. The photographs show that in a stoichiometric mixture a spherical region of hot gas about 1 cm in
diameter is formed very rapidly (in about 15 μs). This region
then expands slowly in a toroidal shape and begins to propagate
as shown in Figure 3. The presence of the cone had no noticeable
effect on how fast the flame developed. The qualitative behavior
of the developing flame followed very closely that which was
observed by Litchfield (4) under similar conditions.

Mass Spectrometric Sampling of Reactants. The principal locations
of the sampling cone were 0.0, 0.2, and 0.5 cm away from the

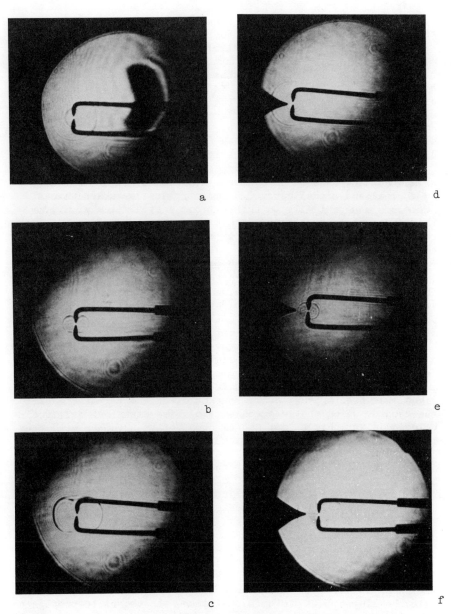

Figure 3.   Ignition kernel and developing flame in the presence and absence of the cone at three different times after the spark. The cone is absent in a, b, and c, and is present in d, e, and f. The pictures were taken at the following times after the spark: a and d, 18 μs; b and e, 64 μs; and c and f, 900 μs.

spark electrodes. The methane signal as a function of time after
the spark is shown in Figure 4 for the cone placed at the electrodes. Results for two equivalence ratios are shown; the ϕ=1.0
mixture ignites and the ϕ=0.35 mixture does not. In the stoichiometric case the methane signal disappears rather slowly. As
a comparison the results in Figure 5a show a much earlier methane
disappearance. In this Figure the cone is located 0.2 cm from
the spark gap. This means that the methane is disappearing 0.2
cm from the gap before it disappears at the spark gap itself.
With the cone located 0.5 cm away from the gap, the methane
disappearance begins at very nearly the same time as it does at
0.2 cm and decreases somewhat more slowly as shown in Figure 6.
While there appears to be a barely noticeable amount of increase
in the methane signal after 0.5 ms in Figure 5a, the methane
decreases monotonically to the post-combustion value at 0.5 cm.
This increase is probably due to unburned methane diffusing into
the 0.2 cm region, although it is clear that the amount is very
small in this case.

     To see how this behavior changes with dilution, the methane
signal for ϕ=1.0 and an oxygen-argon ratio of 0.32 was measured.
The results are shown in Figure 5b. Methane disappearance begins
at about the same time as for O_2/Ar = 0.376, but much more unburned methane is diffusing into this region. A result very
similar to Figure 5b is obtained for O_2/Ar = 0.376, ϕ=0.5 with
the cone located 0.2 and 0.3 cm away from the spark gap.

     Figure 7 shows the oxygen signal as a function of time with
and without time-of-flight chopping for a stoichiometric mixture
and the cone tip located 0.5 cm from the chopper. The chopped
signal has been approximately corrected for the flight time to
the chopper using the measured arrival time distribution. The
velocity distribution (not shown) does not change much with time,
indicating that the apparent stagnation temperature of the sampled gas changes little. This must be due to the thermalization
of the gas as it passes through the cone which is also responsible for low Mach number flow of a room-temperature gas. Subsequent experiments with a 110° cone yield velocity distributions
which are characteristic of the temperature of the sampled gas.
Thermalization of the sampled gas just downstream of the sampling
orifice is an explanation which reconciles these possibly conflicting observations: the low apparent stagnation temperature
of the burned gas and the absence of visible flame distortion
due to the presence of the cone (see Figure 2). Figure 8 gives
the O_2 signal obtained with the sampling cone tip located 0.2 cm
from the electrodes. In both positions, the chopped signal increases abruptly at the time of the spark, and then decays
rapidly within 50-100 μs of the spark. We believe that the
abrupt increase in signal is due to the passage of the shock
wave, although this increase is reflected in the unchopped
signal only at 0.2 cm. We can only speculate that perhaps this
happens because the shock is intense enough at 0.2 cm to be

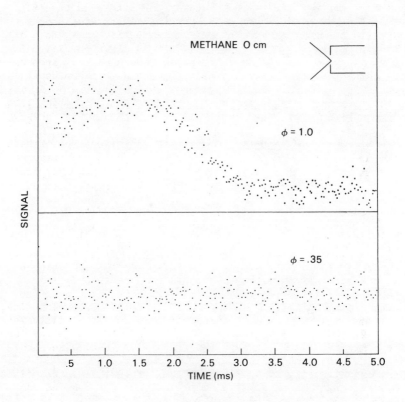

Figure 4.   Methane signal as a function of time for two different
equivalence ratios.   The sampling cone is located at the electrodes.

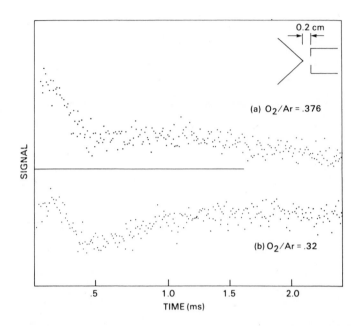

Figure 5. Methane signal for ϕ = 1.0 and two different O_2/Ar ratios. The cone is 0.2 cm from the electrodes.   a, O_2/Ar = 0.376; and b, O_2/Ar = 0.32.

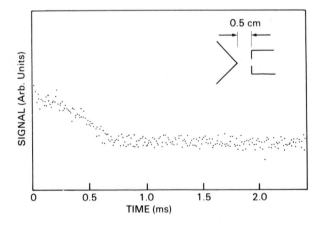

Figure 6. Methane signal for ϕ = 1.0, O_2/Ar = 0.376.   The cone is 0.5 cm from the electrodes.

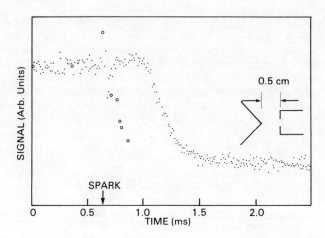

Figure 7.   Oxygen signal for φ = 1.0, O_2/Ar = 0.376, with the cone located 0.5 cm from the electrodes. Both unchopped and chopped signal is shown. The time of the spark is indicated with an arrow. Open circles, chopped; and dots, unchopped.

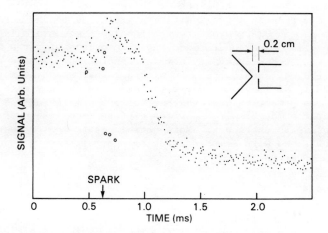

Figure 8.   Oxygen signal for φ = 1.0, O_2/Ar = 0.376, with the cone located 0.2 cm from the electrodes. Chopped and unchopped signals are shown. Open circles, chopped; and dots, unchopped.

noticeable in the unchopped signal but has decayed in strength
at 0.5 cm so that the resulting increase is lost in the averaging over a large range of transit times.

Mass Spectrometric Sampling of Radicals. Methyl radicals and
hydrogen atoms were observed in the spark region. The hydrogen
atom signal with the cone at 0 and 0.2 cm for the stoichiometric
mixture is shown in Figure 9. The H atom signal was undetectable
at 0.5 cm for this mixture. Figure 10 shows the H atom results
for $\phi=0.35$. The signals appear to be stronger than for either
$\phi=1.0$ or 0.5, particularly at 0.5 cm where there is a small but
discernible signal. This is consistent with ignition calculations we have performed using oxygen atom dissociation as the
ignition method. In mixtures that ignite, the initially high
oxygen atom concentration relaxes rapidly to a value characteristic of the flame. Other radicals in the H-O-OH system increase to their characteristic steady-state flame values. In
mixtures which do not ignite, the O atom concentration relaxes
more slowly. An analogous situation occurs if either H or OH
are initially in excess since reactions involving the H_2-O_2
system are very fast at combustion temperatures, driving H, O,
and OH toward their equilibrium values for the H_2-O_2 system at
that temperature. We know that large amounts of H, O, and CH_3
are produced in the spark because CH_3 and H are observed directly
and O can be observed in the absence of methane. In mixtures
that ignite, the concentrations of these radicals should decrease
more rapidly than in a non-igniting mixture due to the more rapid
reactions of the H_2-O_2 system which involve these radicals in the
developing flame and due to more rapid reactions with the fuel.

     As mentioned previously, any interpretation of these H atom
measurements must be looked upon with a rather skeptical eye due
to the non-ideal sampling conditions under which they were obtained. The effect of beam thermalization in altering radical
concentrations may be different for different positions of the
cone relative to the spark. More definitive radical measurements
will be obtained with a more suitable cone. This work is currently in progress.

Discussion

Since the velocity distribution of the sampled oxygen appears to
be approximately independent of the time at which it was sampled,
we can safely assume that this is also the case for methane. If
we assume that the arrival time distribution for methane is approximately the same as that for oxygen except for the difference in
the masses, the time scales of the methane measurement shown in

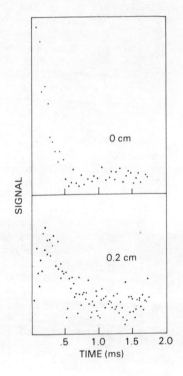

Figure 9, Hydrogen atom signal for
ϕ = 1.0 at two cone locations.

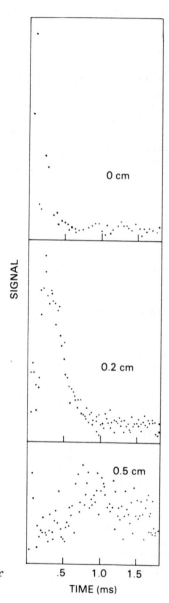

Figure 10. Hydrogen atom signal for
ϕ = 0.35 at three cone locations.

Figures 5a and 6 should be roughly comparable. This means that the methane begins to disappear at 0.5 cm from the cone about 60 μs after it begins to disappear 0.2 cm from the cone. This difference is less than our time resolution, and the result agrees well with the oxygen measurements. Although we have not yet measured the velocity distribution of any components with the cone located at the electrodes, there can be no ambiguity about the observation that methane disappearance begins at 0.2 cm and 0.5 cm before it begins at the electrodes. The only way to conclude from the data that methane disappearance begins at the electrodes at the same time as or sooner than it begins to disappear at 0.5 cm is to assume that the velocity distribution of the sampled methane at 0 cm is characteristic of a much lower apparent stagnation temperature than the methane sampled at 0.5 cm. Since the apparent stagnation temperature of the methane detected at 0.5 cm is near room temperature, this assumption cannot be valid. Therefore, unburned methane must remain in the spark gap after the flame begins to develop. This flame development probably begins at a distance of about 0.2 cm from the electrodes in the stoichiometric mixture. The methane profiles obtained in a more dilute stoichiometric mixture (shown in Figure 5b) and in a mixture where $O_2/Ar=0.376$ and $\phi=0.5$ (not shown) suggest that for these mixtures flame propagation begins at a somewhat greater distance from the spark electrodes.

Lewis and von Elbe (2) and others, in their description of spark ignition, assume that the fuel is burned inside the spark kernel. Apparently this is not the case in the experiment reported here. Results consistent with the presence of unburned fuel in the spark kernel have been reported by Maly and Vogel (5). They found that for a 60 ns breakdown, the center of the spark kernel cooled off very rapidly after the breakdown. They suggested that the shock wave generated by such a fast spark causes a rapid expansion of the gas in the electrode region. This expansion would have little effect on the unburned mixture near the electrodes due to their viscous effect, so the mixture in this boundary layer would be sucked into the rarefied region near the electrodes left behind by the rapidly expanding gas. Our spark is undoubtedly rapid enough (2 μs duration) to produce this same effect because we can observe a shock wave due to the spark in our schlieren photographs. In addition, the cone can cause a boundary layer to be formed which adds to the amount of unburned mixture which can flow into the electrode region. Experiments involving electrodeless ignition, such as by a laser spark, would be useful in verifying whether this phenomenon occurs.

It is not necessary, however, to consider the formation of a viscous boundary layer near the electrodes and sampling cone in order to at least qualitatively explain the oxygen and methane results. We have performed one-dimensional calculations with a code used by us in previous work (6,7) to simulate the ignition

process. We begin the calculation at the time the spark kernel
is fully formed, and assume a temperature distribution centered
in a 4 cm wide container as shown in Figure 11. The position of
maximum temperature gradient is located 0.5 cm on either side of
the center of the container in accord with the outline of the
spark kernel in the schlieren photographs. Oxygen and methane
are dissociated in amounts such that 0.4% of the thermal energy
deposited in each grid region goes into dissociation of oxygen
and 0.4% into methane dissociation at t=0. A heat loss due to
the presence of the spark electrodes is introduced by inserting
an extra term in the energy equation as was done previously (6)
to simulate a cooled crevice. We assume that the electrode gap
is 0.08 cm across and the electrode faces are 0.08 cm wide. The
electrodes are located at the center of the container and are
equidistant from the x axis. The temperature of the electrodes
is the following:

$$T_e(x) = 0.1\ T(x) + 270\ K$$

where $T(x)$ is the temperature of the gas at a point x. This
expression allows the electrodes to increase in temperature as
the gas near the electrodes gets warmer, but at a much slower
rate than the gas.

The resulting time evolution of the O_2 and CH_4 concentrations
at three locations in the container is shown in Figure 12. The
methane and oxygen disappear more rapidly at 0.2 cm than at
either 0.0 or 0.5 cm. If no heat loss is included in the calculations, or if the fraction of heat energy going into dissociation of O_2 and CH_4 is 0.1 rather than 0.004, methane and oxygen
disappearance is very rapid at the center.

We make no claim to have modeled accurately the ignition
process beginning with the formed spark kernel, although that
appears to be possible and will be investigated further. We
performed these calculations to show that at least one phenomenon, which we know must be present in our experiments, can
qualitatively account for the experimental results we have obtained. Viscous boundary layers may also contribute to the
relatively slow burning of methane at x=0.0. We plan experiments
with electrodeless ignition, such as laser sparks, to further
understand this aspect of the ignition process.

Conclusions

The experiments reported here show the chemical effect of heat
loss to the spark electrodes in the ignition process. For rapid
sparks of the type used here, flame development begins about 0.2
cm away from the electrodes rather than at the electrodes or at
the edge of the spark kernel. This new chemical information
provides a description of the spark ignition process which has

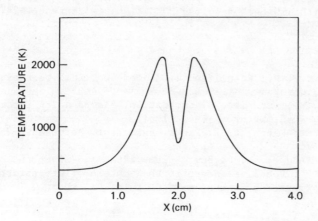

Figure 11.   Initial temperature distribution for the ignition calculation.

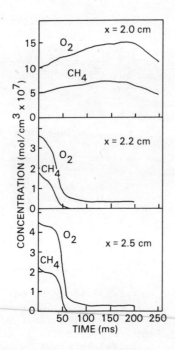

Figure 12.   Oxygen and methane
as a function of time for the
ignition calculation at three
locations.

never been obtained before. As the results obtained are preliminary in nature, however, a more detailed investigation of spark
ignition with improved sampling conditions is currently in progress.

Literature Cited

1. Sloane, T. M.; Ratcliffe, J. W. Combust. Sci. Technol. 1983,
    33, 65.
2. Lewis, B.; von Elbe, G. Combustion, Flames and Explosions of
    Gases; Academic Press Inc., 1961.
3. Yoon, S.; Knuth, E. L. Progr. Astronaut. Aeronaut. 1981,
    74, 867.
4. Litchfield, E. L. Combust. Flame 1961, 5, 235.
5. Maly, R.; Vogel, M. Seventeenth Symposium (International) on
    Combustion; The Combustion Institute: Pittsburgh, 1979,
    p. 821.
6. Sloane, T. M.; Schoene, A. Y. Combust. Flame 1983, 49, 109.
7. Sloane, T. M. Combust. Sci. Technol. 1983, 34, 317.

RECEIVED December 21, 1983

KINETICS AND DYNAMICS

Elementary Combustion Reactions

Laser Photolysis–Laser-Induced Fluorescence Kinetic Studies

FRANK P. TULLY

Sandia National Laboratories, Livermore, CA 94550

A new, laser-based, chemical kinetics technique
has been demonstrated in studies of the reactions
of the hydroxyl radical with ethane and ethylene.
A widely-tunable, quasi-cw, ultraviolet laser
source for exciting transient-species fluorescence
in chemical kinetics experiments has been built
and is described. The reaction between OH and
C_2H_4 is shown to proceed through both OH addition
and H-atom abstraction routes.

Combustion processes are driven by energy-releasing chemical reactions. Detailed knowledge of the chemical kinetics of these
individual reactive steps is required input to combustion models.
For more than a decade, elementary gas-phase reaction kinetics
has been successfully studied with the flash photolysis/resonance
fluorescence technique (1-8). Typically, following broadband
photolysis of a molecular precursor, reactant decays have been
measured under pseudo-first-order kinetic conditions with cw
resonance lamp excitation of free radical fluorescence. Increased
utilization of laser probes in kinetic studies is exemplified by
the recent pulsed-laser photolysis/pulsed-laser-induced fluorescence experiments of McDonald, Lin and coworkers (9-13).
   In the present work, a new kinetics configuration utilizing
a pulsed laser for photolysis and a quasi-cw, ultraviolet laser
for fluorescence excitation has been developed. This technique
combines the best features of the two kinetic methods mentioned
above. Laser photolysis generally permits greater reactant formation specificity than does flashlamp photolysis. Laser-induced
fluorescence detection outperforms resonance fluorescence detection because of its increased fluorescence excitation flux, decreased scattered light signal, and wavelength tunability. Cw
fluorescence excitation is desirable over pulsed fluorescence
excitation due to its freedom from pulse-to-pulse normalization
constraints and, most importantly, because of its efficient duty
cycle and the consequent increased density of points obtainable

0097-6156/84/0249-0225$06.00/0
© 1984 American Chemical Society

in measured molecular concentration versus time profiles. This
high data point density facilitates accurate slope determinations,
readily reveals even subtle deviations from pseudo-first order exponential decays, and offers information on secondary reactions
by carefully mapping such deviations induced by controlled perturbations of the initial reactant conditions.

Many molecular intermediates of importance to combustion and
atmospheric chemistry have primary electronic transitions in the
near ultraviolet region of the electromagnetic spectrum. We have
therefore constructed a widely-tunable, quasi-cw, ultraviolet
laser source for exciting transient-species fluorescence in chemical kinetics experiments. As summarized in Figure 1, a mode-
locked Ar^+ laser operating at 514.5 nm synchronously pumps an
extended-cavity dye laser, producing, with various dyes, tunable
radiation from 540 nm to 900 nm. The dye laser fundamental output consists of a train of pulses of 3–6 nJ energy and 8–10 ps
duration at a repetition rate of 246 MHz. This output is
frequency doubled using temperature- and angle-tuned second harmonic generation crystals. The ultraviolet laser radiation produced in this process is then used to excite fluorescence in the
reactive radicals of interest.

Three features of this laser source merit further discussion.
First, in a typical kinetic experiment, the 1/e chemical lifetime
of the photolytically produced radicals varies between 0.2 and
25 ms, a representative mean being $t_{1/e}$ = 2 ms. For statistical
reasons, one desires to collect a minimum of 20 concentration
versus time data points per 1/e concentration decay period. For
multichannel scaling detection, these typical kinetic conditions
imply a maximum dwell period per channel of 100 μs. The ultraviolet laser source described above emits 2.5×10^4 pulses per
100 μs interval; thus, relative to chemical decays, this rapidly
pulsed source is viewed by the experiment as a cw excitation
probe. Second, given that a pulsed initiation/cw detection
kinetics configuration is desired, one may ask why a cw laser
source is not used. The rationale here is that the visible-to-
ultraviolet conversion efficiency is much higher when the quasi-
cw source rather than a cw source is used. Frequency doubling
efficiency varies in proportion to the fundamental peak power
density present in the second harmonic generation crystal,
$(P_{2\omega}/P_\omega) \alpha P_\omega$. Table I lists typical pulse repetition rates,
fundamental peak power densities and frequency doubling effi-
ciencies obtainable with various visible laser sources. For the
cw and quasi-cw dye laser sources, peak power densities are estimated assuming that 1.0 watt of visible radiation is focused
to a 50 μm spot within the frequency doubling crystal. Because
the beam energy is bunched into short duration pulse packets
with the quasi-cw source, the obtainable focused peak power density and the resultant second harmonic generation efficiency are
much larger with this source than with a cw dye laser source.
From 1.0 watt of dye laser radiation at 616 nm, for example, we

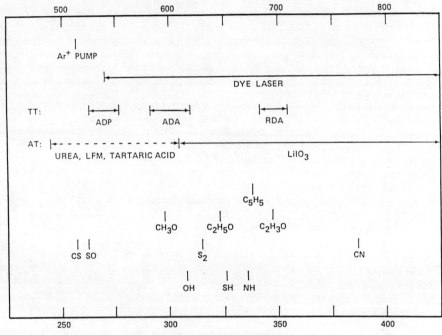

Figure 1.   Laser-induced fluorescence detection of UV-absorbing
free radicals. The vertical lines denote the wavelengths that
are most useful in fluorescence excitation.

Table I.   Second Harmonic Generation from Visible Dye Lasers

	Pulse Repetition Rate (Hz)	Peak Power Density (MW/cm^2)	Second Harmonic Generation Efficiency (%)
Nd/YAG-pumped dye laser	30	50, unfocused	10-25
cw dye laser	cw	0.05, focused	0.1
quasi-cw dye laser	2.46 x 10^8	20, focused	8

obtain average ultraviolet laser powers of 80 mW and 1 mW upon frequency doubling the quasi-cw and cw fundamental beams, respectively. The high ultraviolet flux so obtained with the quasi-cw laser source will permit efficient study of many previously unobservable chemical processes. Finally, the uncertainty principle dictates that short duration laser pulses have wide spectral bandwidths. We measure FWHM laser linewidths of $\approx$ 50 GHz for the fundamental beam. This width is much larger than a typical Doppler-broadened absorption line in a diatomic molecule, and, as described below for OH detection, absorption line coincidences must be located and exploited to optimize the laser-induced fluorescence detection efficiency of diatomic radicals with this quasi-cw, ultraviolet source. This constraint largely disappears when monitoring polyatomic radicals, as the separations between their ro-vibronic transition lines are comparable to Doppler linewidths, thereby making all of the source ultraviolet radiation absorbable by the radical. These considerations are discussed in detail by Inoue et al (14) in their comparison of the laser-induced fluorescence spectroscopies of the molecular homologs OH and CH_3O.

The application of the quasi-cw, ultraviolet laser source to kinetic studies was demonstrated in the laser photolysis/laser-induced fluorescence experiments shown schematically in Figure 2. Chemical reactions were initiated by 193-nm photolysis of N_2O in N_2O/H_2O/hydrocarbon/helium gas mixtures. The $O(^1D)$ atoms formed by photodissociation rapidly converted to OH through reaction with H_2O, and time-resolved OH concentrations were measured as functions of hydrocarbon number density by laser-induced fluorescence. Hydroxyl radical fluorescence was excited by pumping the nearly coincident $P_1 1$, $Q_1 3$, and $Q_1 3'$ (0,0) band $X^2\Pi \rightarrow A^2\Sigma^+$ transitions at 308.16 nm, (15) and radiation emanating from the reaction volume in a downward direction was skimmed by black-anodized collimators, focused by quartz lenses, selected by a bandpass filter (308.3 nm peak, 8 nm FWHM), and detected by an RCA 8850 photomultiplier operating in the photon-counting mode.

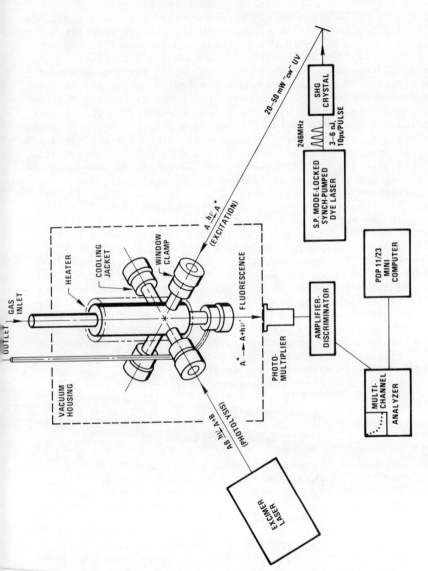

Figure 2. Schematic diagram of the laser photolysis/laser-induced fluorescence chemical kinetics apparatus. (Reproduced with permission from Ref. 20. Copyright 1983, North-Holland.)

The photomultiplier output pulses were amplified, discriminated, and fed into a multichannel scaler, and OH fluorescence decays were signal averaged over 25-250 excimer laser shots.

The first kinetics experiments performed with this apparatus dealt with the abstraction of hydrogen atoms by OH from methane and ethane, OH + RH$\longrightarrow$H$_2$O + R. Reliable rate coefficient data for these reactions had previously been obtained in flash photolysis/resonance fluorescence studies, (7,16) and agreement with these published data served as a required check on the performance of the new kinetics configuration. The results for the reaction between OH and C$_2$H$_6$ are shown in Figure 3. The rate coefficients measured in the present work all fall about 5% below those of Ref. 16; such agreement is well within the estimated 10% accuracy limits of the two studies. In both sets of experiments, the OH excitation and fluorescence wavelengths were resonant, and optimizing the [OH]-time profiles required maximizing the detected fluorescence signal (S) while minimizing the detected scattered light background (B). For similar values of $[OH]_{t=0}$, the ratio S/B in the present work exceeded that of Ref. 16 by more than an order of magnitude. Two factors contributed to this marked improvement in S/B. First, the absorbable photon flux generated with the quasi-cw, ultraviolet laser source exceeded that from an OH resonance lamp by a factor of 2-3. Second, and most significant, the detected scattered light signal from this collimated laser source was 10-25 times less than that typically obtained with a volume-source, OH resonance lamp (8). Further improvements in S/B are expected in future experiments in which OH fluorescence will be excited by single-frequency ultraviolet laser radiation obtained by intracavity frequency-doubling an actively stabilized cw ring dye laser.

Encouraged by these results, we began to study hydroxyl radical reactions for which only limited kinetic information is available. A detailed investigation of the reaction

$$OH + C_2H_4 \xrightarrow{k_1} Products \tag{1}$$

is in progress. At present, kinetic measurements have been made at 600 Torr helium throughout the temperature range 291-796 K, and at 291 K over the pressure range 50-600 Torr helium. Absolute reaction rate coefficients k_1 were determined, or, in some cases, approximated, as described below.

We carried out all experiments under pseudo-first-order kinetic conditions with $[OH] \ll [C_2H_4]$. Excluding secondary reactions that significantly deplete or reform OH, [OH] varied exponentially with time:

$$[OH]_t = [OH]_o \exp -(k_1[C_2H_4] + k_d)t = [OH]_o \exp(-k't)$$

where k' is the measured pseudo-first-order decay rate, k_1 is the bimolecular rate coefficient for Reaction (1), $[C_2H_4]$ is

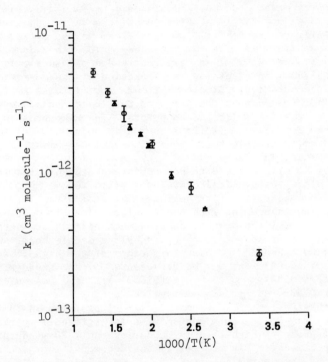

Figure 3. Arrhenius plot of rate coefficient data for the reaction OH + C_2H_6 → H_2O + C_2H_5. O , Ref. 16, flash photolysis/resonance fluorescence; Δ, this work, laser photolysis/laser-induced fluorescence. Error specifications are ±2 precision limits in both cases.

the (constant) ethylene concentration, and k_d is the first-order
rate constant for OH removal in the absence of $[C_2H_4]$ due to diffusion from the reaction volume and to reaction with background
impurities. We observed exponential [OH] decays, such as that
shown as ln [OH] versus time in Figure 4a, for all experiments
at T = 291 and 361.5 K. However, as we varied T from 438 to 666
K, non-exponential features in the [OH] profiles became increasingly apparent. A typical [OH] profile obtained at 591 K is
plotted for comparison in Figure 4b. For exponential [OH] profiles, -k' was equated to the calculated least-squares slope of
the decay taken over at least a factor of ten variation in [OH].
When analyzing nonexponential [OH] profiles, we estimated -k'
from the steep initial slope of the decay. In either case, the
k' values obtained at a given temperature and pressure were
plotted, as shown in Figure 5, as a function of the corresponding
ethylene concentration. Bimolecular rate coefficients $k_1(T,P)$
were then determined from the slope of the least-squares straight
line through the $([C_2H_4],k')$ data points. The high-pressure-
limited rate coefficients $k_1(T)$ measured in this work and in
previous studies are plotted, along with various summary analytic
representations, in Figure 6.

At 291 K, k_1 was found to be pressure-dependent, and it
reached a high-pressure-limited value of $(8.47\pm0.24) \times 10^{-12}$ cm^3
molecule^{-1} s^{-1} above 400 Torr helium. This value for k_1(291 K)
is in excellent agreement with the results of previous studies
(17-19). From 291-438 K, the reaction mechanism is dominated by
electrophilic addition of OH to the ethylene double bond, and
the temperature dependence over this range of the high-pressure-
limited rate coefficient may be represented by the expression
$k_1(T) = (1.74\pm0.53) \times 10^{-12}$ exp $(918\pm214)/RT$ cm^3 molecule^{-1} s^{-1},
where quoted errors represent $\pm2\sigma$ values and $\sigma_A = A\sigma_{lnA}$ (20,21).

Our observation of nonexponential [OH] profiles in the
temperature range 438-666 K can only be interpreted in terms of
a chemical process which reforms OH during the 1-20 ms duration
of the experiment. This process is the decomposition back to
reactants of the thermalized adduct HOC_2H_4. Indeed, at 591 K,
we observed that at very long times the [OH] decays again became
exponential with a slope of $-k_d$ (see Figure 4b). This situation
results only when the OH + C_2H_4 + M $\rightleftharpoons$ HOC_2H_4 + M reaction has
established dynamic equilibrium, with the thermalized adduct
serving, in effect, as a temporary sink for OH. The rate coefficient data derived from nonexponential [OH] profiles must
thus be considered approximate, and they are included in Figure
6 only to show the decreasing trend in "net" reactivity between
OH and C_2H_4 with increasing temperature. These results and their
interpretation are entirely analogous to those obtained in studies of hydroxyl radical addition to aromatic hydrocarbons (22-25).

We have also made very preliminary kinetic measurements on
Reaction (1) at T = 704 and 796 K. The [OH] profiles collected
at these temperatures show a lower degree of nonexponential

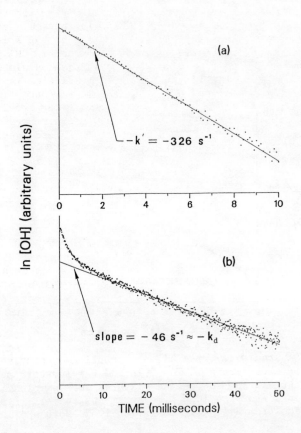

Figure 4. Typical [OH] concentration profiles obtained in kinetic measurements of the reaction OH + C_2H_4 ⟶ products; a, T = 291K, P = 600 torr helium, and $[C_2H_4]$ = 3.40 x 10^{13} molecule cm^{-3}; b, T = 591K, P = 600 torr helium, and $[C_2H_4]$ = 2.78 x 10^{14} molecule cm^{-3}. (Reproduced with permission from Ref. 20. Copyright 1983, North-Holland.)

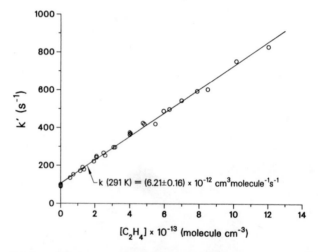

Figure 5. Measured decay rate k' as a function of ethylene concentration for experiments at T = 291K and P = 100 torr helium. Three different C_2H_4/He source gas mixtures were sampled. The estimated accuracy of the $[C_2H_4]$ measurements is 5%. (Reproduced with permission from Ref. 20. Copyright 1983, North-Holland.)

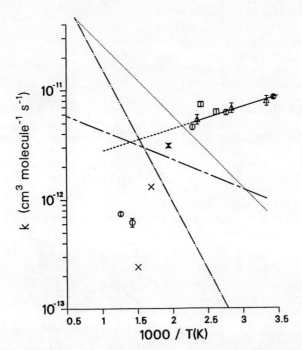

Figure 6. Arrhenius plot for the reaction OH + C_2H_4 $\longrightarrow$ products: O, this work, exponential [OH] decays; X, this work, nonexponential [OH] decays; $\triangle$, Ref. 17; $\square$, Ref. 21; $-----$, Ref. 20; $\cdots\cdots$, Ref. 26; $\overline{\quad}\cdot\overline{\quad}\cdot$, Ref. 27; and $\overline{\quad}-\overline{\quad}-$, Ref. 28.

236   CHEMISTRY OF COMBUSTION PROCESSES

behavior than those measured in the range 515–666 K. However,
k' values at fixed ethylene concentrations varied somewhat when
photolysis pulse energies were changed; we have not yet identified the source of this effect. Nevertheless, semi-quantitative
estimates of k_1(704–796 K) may be derived from these measurements, and the reaction rate coefficient appears to grow with
increasing temperature over this range. We believe this increase
in k_1(T) to be caused by the onset of the hydrogen abstraction
channel for Reaction (1), $OH + C_2H_4 \longrightarrow H_2O + C_2H_3$. Two-
parameter expressions for this abstraction channel rate coefficient have previously been derived from complex, high-temperature
kinetic studies in three reviews (26–28). As seen in Figure 6,
our preliminary measurements yield poor agreement with these
recommendations. Further direct, "single-reaction" studies of
this abstraction process will be needed to clarify these dis-
crepancies.

In summary, we have demonstrated that photolysis/fluores-
cence chemical kinetics techniques must exploit ongoing advances
in laser technology. A highly-sensitive, quasi-cw, ultraviolet
laser source was constructed and used in definitive chemical
kinetics experiments. OH–ethylene reactions have been charac-
terized in terms of OH addition and hydrogen atom abstraction
channels, and questions have been raised concerning both the
importance of the addition process and the accuracy of presently
recommended kinetic parameters for the abstraction process at
combustion temperatures.

Acknowledgments

This research was supported by the Office of Basic Energy
Sciences, U. S. Department of Energy. The author gratefully
acknowledges helpful discussions with J. E. M. Goldsmith,
J. S. Binkley, M. L. Koszykowski, C. F. Melius and R. E. Palmer.
He also wishes to thank R. D. Gay, S. C. Gray, A. R. Van Hook
and G. D. Cosgrove for their contributions to this work.

Literature Cited

1. Braun, W.; Lenzi, M. Discussions Faraday Soc. 1967, 44, 252.
2. Kurylo, M. J.; Peterson, N. C.; Braun, W. J. Chem. Phys.
    1970, 53, 2776.
3. Davis, D. D.; Huie, R. E.; Herron, J. T.; Kurylo, M. J.;
    Braun, W. J. Chem. Phys. 1972, 56, 4868.
4. Stuhl, F.; Niki, H. J. Chem. Phys. 1972, 57, 3671.
5. Davis, D. D.; Fischer, S.; Schiff, R. J. Chem. Phys. 1974,
    59, 628.
6. Atkinson, R.; Hansen, D. A.; Pitts, J. N., Jr. J. Chem.
    Phys. 1975, 62, 3284.
7. Tully, F. P.; Ravishankara, A. R. J. Phys. Chem. 1980, 84,
    3126.

8. Ravishankara, A. R.; Nicovich, J. M.; Thompson, R. L.;
    Tully, F. P. J. Phys. Chem. 1981, 85, 2498.
9. McDonald, J. R.; Miller, R. G.; Baronavski, A. P. Chem.
    Phys. 1980, 30, 133.
10. Lin, M. C.; McDonald, J. R. in "Reactive Intermediates in
    the Gas Phase"; Setser, D. W., Ed.; Academic: New York,
    1979; p. 233.
11. Sanders, N.; Butler, J. E.; Pasternack, L. R.; McDonald,
    J. R. Chem. Phys. 1980, 48, 203.
12. Butler, J. E.; Fleming, J. W.; Goss, L. P.; Lin, M. C.
    Chem. Phys. 1981, 56, 355.
13. Nelson, H. H.; McDonald, J. R. J. Phys. Chem. 1982, 86,
    1242.
14. Inoue, G.; Akimoto, H.; Okuda, M. J. Chem. Phys. 1980,
    72, 1769.
15. Dieke, G. H.; Crosswhite, H. M. J. Quantum Spectry. Rad.
    Transfer 1962, 2, 97.
16. Tully, F. P.; Ravishankara, A. R.; Carr, K. Intern. J.
    Chem. Kinet. 1983, 15, 1111.
17. Atkinson, R.; Perry, R. A.; Pitts, J. N., Jr. J. Chem.
    Phys. 1977, 66, 1197.
18. Lloyd, A. C.; Darnall, K. R.; Winer, A. M.; Pitts, J. N.,
    Jr. J. Phys. Chem. 1976, 80, 789.
19. Cox, R. A. Intern. J. Chem. Kinetics Symp. 1975, 1, 379.
20. Tully, F. P. Chem. Phys. Lett. 1983, 96, 148.
21. Gordon, S.; Mulac, W. A. Intern J. Chem. Kinetics Symp.
    1975, 1, 289.
22. Perry, R. A.; Atkinson, R.; Pitts, J. N., Jr. J. Phys.
    Chem. 1977, 81, 296.
23. Perry, R. A.; Atkinson, R.; Pitts, J. N., Jr. J. Phys.
    Chem. 1977, 81, 1607.
24. Tully, F. P.; Ravishankara, A. R.; Thompson, R. L.;
    Nicovich, J. M.; Shah, R. C.; Kreutter, N. M.; Wine, P. H.
    J. Phys. Chem. 1981, 85, 2262.
25. Nicovich, J. M.; Thompson, R. L.; Ravishankara, A. R.
    J. Phys. Chem. 1981, 85, 2913.
26. Warnatz, J. in "Chemistry of Combustion Reactions";
    Gardiner, W. C., Jr., Ed.; Springer-Verlag: New York, 1983;
    Chap. 5.
27. Westley, F. "Table of Recommended Rate Constants for
    Chemical Reactions Occurring in Combustion"; NBSIR 79-1941,
    November 1979.
28. Westbrook, C. K.; Dryer, F. L. "Chemical Kinetics Modeling
    of Hydrocarbon Combustion"; Lawrence Livermore National
    Laboratory Report No. UCRL-88651; February, 1983.

RECEIVED November 30, 1983

Reaction Rate of OH and C_2H_2 near 1100 Kelvin

Laser Pyrolysis–Laser Fluorescence Measurement

PAUL W. FAIRCHILD[1], GREGORY P. SMITH, and DAVID R. CROSLEY

SRI International, Menlo Park, CA 94025

A laser pyrolysis/laser fluorescence apparatus has been used to measure the rate constant for the reaction of hydroxyl radicals with acetylene at temperatures near 1100K. A pulsed CO_2 laser rapidly heats a mixture of SF_6, N_2, and H_2O_2, pyrolyzing the peroxide to form OH. The radicals are detected by laser-induced fluorescence, produced by a dye laser fired at a variable time delay following the infrared laser. Addition of C_2H_2 causes a decrease in the OH signal with time, permitting a measurement of the rate constant. At 1100 ± 50K a value of $3 \times 10^{-13} cm^3 sec^{-1}$ was obtained, with no significant pressure dependence over the range 10-100 torr. A Troe-type calculation has been carried out for the pressure-dependent addition channel known to occur for this reaction at lower temperature, and shows a temperature dependence of the fall-off such that the rate channel should be slow at 1100K. Thus our results and these previous experiments are consistent, and a direct reaction channel important above 1000K is indicated. This combined temperature/pressure dependence must be included in combustion models incorporating detailed chemical kinetics.

As exemplified by several papers within this symposium, computer models of combustion processes which incorporate extensive and detailed chemical kinetics are now feasible. Such calculations are capable of treating systems varying widely in time

[1]Current address: TRW, Space and Technology Group, Redondo Beach, CA 90278

scale - stable flames through detonations - and over a large range
of temperature and pressure. The input to these models consists
of mechanistic and rate constant data for the reactions involved;
for at least some of the key reactions these rate constants must
be well determined from independent experiments. Particularly
important is knowledge of the rate constants at intermediate and
high temperatures and, in cases where three-body addition
mechanisms can contribute, at varying pressures at each temperature.   Simple extrapolation of room temperature values is often
inadequate for these purposes, while measurements directly at high
temperatures, as in flames themselves, are often complicated by
the occurrence of other reactions as well.

We have developed a new technique of laser pyrolysis/laser
fluorescence, or LP/LF (1), designed to furnish direct measurement
of rate constants of reactions involving free radicals at elevated
temperatures (800-1400K).   A pulsed CO_2 laser is used to heat a
sample containing a precursor that pyrolyzes to form radicals.
These radicals are then detected using laser-induced fluorescence
(LIF).   The measurement of the radical removal rates in the
presence of added reactant then yields the rate constant for the
selected conditions of temperature (T) and pressure (P).

In the present study we have applied the LP/LF method to the
measurement of the rate constant for the $OH + C_2H_2$ reaction, at
T ~ 1100K and P between 10 and 100 torr. This reaction may be of
crucial importance in the mechanism of soot formation, a topic
also treated in several papers within this symposium. Previous
direct measurements have been performed in the range T = 230 to
420K, where a distinct pressure dependence indicated the dominance
of an addition mechanism.   We find little or no pressure
dependence at the elevated temperature and an overall rate
constant below the room temperature value.

A theoretical description of the addition reaction, based on
Troe's formulation (2) of unimolecular reaction rate theory, has
been constructed to address the question of the consistency of our
results and the earlier low temperature measurements.   These
calculations show a dramatic combined temperature and pressure
dependence of this rate constant which must be included when this
reaction is incorporated into models of combustion chemistry.
These results illustrate the need to combine individual
experimental data with a theoretical overview in order to obtain a
description valid over the range of T and P likely encountered in
combustion systems.

Laser Pyrolysis/Laser Fluorescence

The LP/LF technique utilizes a pulsed CO_2 laser as the heating
source and a pulsed dye laser, delayed in time with respect to the
CO_2 laser, as the probe laser for detection of the radicals. The
experimental set-up is shown in Figure 1. A gas mixture containing an infrared absorber (SF_6), a bath gas (N_2), a radical

precursor (H_2O/H_2O_2), and a reactant (C_2H_2) flows through the cell and is irradiated by the CO_2 laser. The SF_6 absorbs the CO_2 laser radiation and collisionally transfers this energy to the surrounding gas. The temperature of the heated mixture is chosen by the CO_2 laser fluence and the SF_6 pressure used. Once heated, thermal decomposition of the peroxide produces hydroxyl radicals. The frequency doubled dye laser is tuned to an absorption line of the OH (typically the $P_1 6$ line for the (0,0) band of the $A^2\Sigma^+$-$X^2\Pi_i$ transition), producing fluorescence which is detected through a small monochromator and averaged with a gated boxcar integrator. Variation of the delay between the lasers permits the relative OH density to be mapped as as function of time following the initial heating pulse. By tuning the dye laser through a series of rotational absorption lines, the population distribution among ground state rotational levels and hence the temperature can be determined at each delay time.

The dye laser beam is 2 mm in diameter and the fluorescence is focussed onto a 2 mm slit oriented perpendicular to the beam. This provides point-type probing of the heated gas, and permits us to select a particular spatial region for performing the kinetics experiments.

Following the initial heating pulse, a series of gas dynamic processes take place, altering the environment in the heated region, which is cylindrical with a 1 cm diameter. A shock wave proceeds outward from the heated region while a rarefaction (expansion) wave propagates inward. Upon reaching the center of the heated cylinder, this expansion wave reflects and proceeds outward; the net result is a two-stage cooling and an expansion of the heated gas. This cooling process takes place in ~ 30 μsec, that is, the transit time for the expansion wave across the heated region at the speed of sound at the elevated temperature. The typical cooling of 200–300K effectively quenches the high activation energy radical production reaction. After this time, there continue to be minor reflected and refracted waves near the center of the cell, but the density and temperature remain constant to within about 5% over the 100 μsec time period in which the reaction rates are measured. Further cooling occurs slowly (< 1K/μsec) by thermal conductivity across the interface between the initially heated and unheated regions, but does not affect results on the time scale of the present experiments.

Although we directly measure the temperature in each experiment, we must rely on a computer calculation (3) of the temperature-density relationship produced by the gas dynamic processes. In order to verify that the code is properly describing the situation, we have performed a series of measurements of the OH density and temperature as a function of time (1,4). In Figure 2 is shown the evolution of T following the CO_2 laser pulse for two different gas mixtures, pure SF_6 and $SF_6 + N_2$. In the latter case, which has a higher heat capacity ratio, a larger cooling is observed. These measurements,

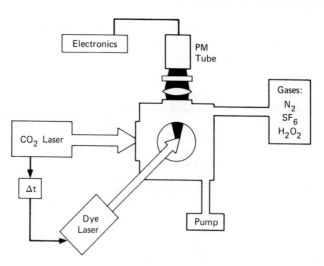

Figure 1. Schematic diagram of the experiment. The time delay
Δt is actually controlled by an external circuit that fires the
CO_2 and YAG laser sequentially. The fluorescence is focused onto
the slit of a monochromator, detected by a photomultiplier tube,
and averaged with a gated boxcar integrator. (Reproduced with
permission from Ref. 1. Copyright 1982, The Combustion Institute.)

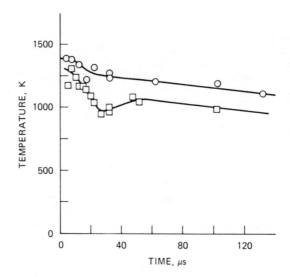

Figure 2. Measured temperature as a function of time, at cell
center. Circles, 100 torr SF_6 only, $T_{initial}$ = 1400K. Note the
low value of γ produces less cooling by the expansion wave. Squares,
100 torr SF_6, 30 torr N_2, $T_{initial}$ = 1300K. (Reproduced with
permission from Ref. 1. Copyright 1982, The Combustion Institute.)

performed at the center of the cell where the influence of the gas dynamics and potential fluctuations are greatest, are in qualititative agreement with the code calculations. In particular, the overcooling of the mixture at about 20 μsec appears in both the computation and the experiment. In Figure 3 is shown the signal level as a function of delay time at the center of the cell. A narrow gate is used on the boxcar so that the signal is proportional to the OH density and is unaffected by quenching collisions (4). The rise in the fluorescence reflects the production of OH from the pyrolysis of the H$_2$O$_2$. The short induction time and linear rise of the OH signal indicates that the decomposition is time-independent and that the energy transfer from the SF$_6$ is fast. This has been verified in subsequent experiments in which the rotational temperature of the OH measured by 1-1 and 0-0 excitation are found to be the same, within experimental error. The solid line in Fig. 3 shows the density profile predicted by the code at the center of the cell. The decline and final value, at times > 50 μsec, follow well. The disagreement between computation and experiment for the oscillation near 30 μsec is caused at least partially because the laser beam averages over the central 20% of the cell, while the code predictions are for the exact center, where such oscillations are most severe.

Halfway between the cell center and the edge of the heated region, the code predicts only minor oscillations and fluctuations in the density, and very little initial overcooling. This is the region chosen for the kinetics measurements. Confidence in our ability to correctly predict the density in this region is obtained from the previous measurements (1) of the OH+CH$_4$ rate constant between 800 and 1400K, which show, within the scatter of individual points of ±30%, very good agreement with the values expected from an extrapolation of results up to 1050K from other investigators.

The rate constants are then determined by measuring the time dependence of the OH density, first in the absence and then in the presence of a known amount of added reactant. The measurements begin at 30-40 μsec delay, after the expansion waves have settled. From the logarithm of the ratio versus time, the rate is obtained; the slope of these rates vs the reactant pressure then furnishes the rate constant for the given conditions. Reactant concentrations are chosen large enough to effect an observable decay during the 100 μs measurement time interval, and are also large enough compared to OH to ensure that pseudo-first-order kinetics is valid.

One feature of the LP/LF method is the rapid heating, compared to conventional heated flows. This permits reactions to be measured under conditions when one or another of the reactions might pyrolyze during the much longer residence times needed for thermalization in a flow cell. On the other hand, the temperature (measured for each run by the rotational excitation scans) is not

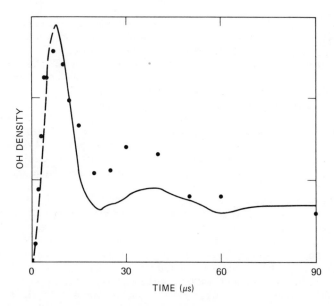

Figure 3. Narrow gate experiment measuring the density of OH as a
function of time after the CO_2 heating pulse, at the center of the
pyrolysis cell. Points, experimental results; solid line, predictions from computer calculation; and dashed line, pyrolysis rate of
H_2O_2 at the initial temperature. (Reproduced with permission from
Ref. 4. Copyright 1983, Journal of Chemical Physics.)

known as accurately as in flow tubes; the precision is ± 50K here. This does not constitute a significant source of error in the present experiments, where the temperature dependence of the reaction rate is gentle, but may for other cases. Also, the subsequent thermal conductivity cooling limits the range of time over which measurements can be made.

Experimental results

Data for the decay of the OH density for several pressures of C_2H_2 at a particular T and P are shown in Fig. 4, as well as the plot from which the rate constant for these conditions is determined. The scatter exhibited is in accord with our estimates of 10-20% run-to-run error. A series of such measurements was made for bath gas pressures ranging from 10 to 100 torr. Typical bath gas mixtures contained equal amounts of N_2 and SF_6. The results are given in Fig. 5. As seen in the figure, there are two sets at T = 1070 ± 50K and 1180 ± 40K for which the results are indistinguishable within the error bars. The dashed line is a least-squares fit to the rate constants as a function of pressure. The fitted slope is consistent with no pressure variation, given the spread in the results. We conclude that at this temperature, ~ 100K, the OH + C_2H_2 reaction is independent of pressure and proceeds through a direct channel such as abstraction. Within 2σ error bars, at least 65% of the reaction proceeds by this channel at 100 torr. Assuming only a pressure independent mechanism, these results give k =(2.6 ± 0.3) x 10^{-13} cm^3s^{-1} at 1140K average temperature.

Preliminary runs carried out at ~ 900K indicate that at this cooler temperature there does exist a significant pressure-dependent component (5). The error bars shown in Fig. 5 are average deviations from the mean of the 3-5 measurements made at each T and P using various C_2H_2 partial pressures. Error estimates derived from data scatter, laser fluctuations (monitored at less than 5%) and flow reading accuracy, and uncertainties in the density resulting from both error in the temperature measurements and confidence in the code calculations give similar values. Error estimates are discussed in more detail in Refs. (1, 4, and 5). The scatter of the OH+CH_4 results from expected values agreed well with the error bars estimated separately in this way. The uncertainty of ~ 50K in each temperature is directly obtained from the scatter in the Boltzmann plots used to obtain T from the rotational populations.

Discussion

The reaction of OH with C_2H_2 appears to have two primary reaction pathways. Measurements made in the 230-430K temperature range (6, 7, 8) show an unambiguous pressure dependence. Thus at these cooler temperatures the reaction is dominated by an addition channel

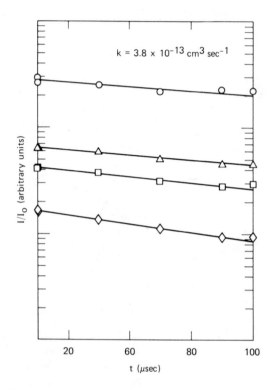

Figure 4a. Ratio of OH LIF signal with added C_2H_2 to that with no added C_2H_2 as a function of delay time after the CO_2 laser pulse. T = 1170K and P_{TOT} = 32 torr. Circles, 1.02 torr C_2H_2; triangles, 1.85 torr C_2H_2; squares, 2.27 torr C_2H_2; and diamonds, 2.53 torr C_2H_2.

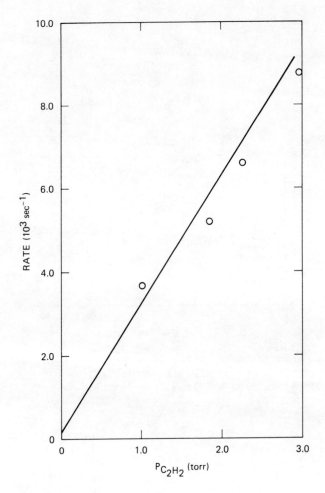

Figure 4b. Plot of decay rate as a function of added C_2H_2 pressure. $k = 3.8 \times 10^{-13}$ cm^3 sec^{-1}.

OH + C_2H_2 P DEPENDENCE

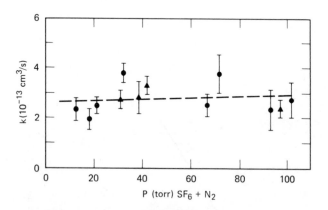

Figure 5. Measured value of the OH + C_2H_2 rate constant as a function of total pressure of SF_6 + N_2. Circles, 1180 + 40K; and triangles, 1070 $\pm$ 50K. Dashed line is a least-squares fit to the rate constants.

$$\text{OH} + C_2H_2 + M \underset{k_{-1}}{\overset{k_1}{\rightleftharpoons}} C_2H_2OH + M \tag{1}$$

in which the initially energetic adduct $C_2H_2OH^*$ is stabilized by collision with some other molecule. At higher temperatures the reaction likely proceeds by a direct mechanism, probably abstraction

$$\text{OH} + C_2H_2 \rightarrow C_2H + H_2O \tag{2}$$

although previous high temperature measurements (9, 10) were made only in a flame, each at a single pressure, and must be strictly considered inferential regarding the reaction channel and rate. The present experimental results show clearly, however, that a direct, pressure-independent channel is the main one in the neighborhood of 1100K.

Our results taken together with these lower temperature measurements (6, 7, 8) raise questions which we have attempted to answer using unimolecular and bimolecular reaction rate theory. The first is, simply, are the results at low and high temperature consistent? Further, what is the expected behavior in the intermediate region and under what conditions does the direct or addition channel dominate?

We have calculated the addition channel rate constant using the RRKM approach to unimolecular reaction rate theory, as formulated by Troe (2) to match RRKM results with a simpler computational approach. The pressure dependence of the addition reaction (1) can be simply decribed by a Lindemann-Hinshelwood mechanism, written most conveniently in the direction of decomposition of the stable adduct:

$$C_2H_2OH + M \underset{k_{-3}}{\overset{k_3}{\rightleftharpoons}} C_2H_2OH^* + M \tag{3}$$

$$C_2H_2OH^* \overset{k_4}{\rightarrow} C_2H_2 + OH \tag{4}$$

The * indicates a molecule having sufficient energy for decomposition. Then

$$k_{-1} = \frac{k_3 k_4 [M]}{k_{-3}[M] + k_4} = \frac{k_4 K_3}{(1 + k_4/k_{-3}[M])}$$

$$\equiv k_\infty (1 + k_\infty/k_0[M])^{-1} \tag{5}$$

and the rate constant for the addition reaction is given by

$$k_1 = K_1 k_{-1} \tag{6}$$

where the K_i are equilibrium constants. Eqn. (5) provides an inadequate description of the kinetics of reaction (1). The reason is that the rate constants k_{-3} and k_4, viewed microscopically, depend on the energy of the hot adduct $C_2H_2OH^*$, and this must be accounted for in a proper description. Here lies also the physical reason for the temperature dependence of the pressure fall-off region of the rate constant, that is, where the reaction is between second and third order overall. At higher temperature, the $C_2H_2OH^*$ will span a larger range of energy and have a higher average energy. Thus reaction (4) will be faster while reaction (3) will require more collisions to stabilize the more energetic adduct.

The correct approach is the subject of RRKM theory; we use the computationally simpler procedure as given by Troe (2). He writes k_{-1} in the form of a correction to the Lindemann-Hinshelwood expression

$$\log k_{-1} = \log [k_\infty/(1 + k_o[M])]$$

$$+ \log F_T/[1 - \{\log(k_\infty/k_o[M])\}^2] \qquad (7)$$

The factor F_T is a modification of reduced Kassel integrals designed to replicate full RRKM results and is calculable from known or estimated thermochemical/structural properties of the adduct. k_∞ is obtained from transition state theory, using the partition function of the activated complex and $E^{\#}$, the barrier to recombination. These are determined from the size and the temperature dependence of the experimental high pressure limiting rate constants. k_o is given by an expression involving a Lennard-Jones collision rate constant, an integral involving the density of states for the adduct, approximated here by the procedure given in (2), and a parameter β which represents the efficiency of energy transfer in an average collision between $C_2H_2OH^*$ and M.

The details of the computational procedures are given in (2), and the choice of the parameters used is discussed in (5). Briefly, the adduct structure is estimated with good confidence by analogy to similar molecules, and $E^{\#}$ is determined to be 1.2 kcal/mole, by the best fit to the temperature dependence of k_∞ as determined by Michael et al. (7) over the range 227-413K. The remaining parameters which are then adjusted to produce the best fit are the frequencies of the CCO deformation and OH torsion mode of the adduct in the transition state, the thermodynamic stability of the adduct, and the value of β. The values used for these parameters must be constrained to be physically reasonable. The final values, 205 cm^{-1} for each of the two vibrations, an adduct stable by 34 kcal/mole, and an average energy transferred per collision of 0.36 kcal/mole, fulfill this criterion.

The results of the fits are compared in Fig. 6 with the pressure dependence of k_1 for the five temperatures investigated

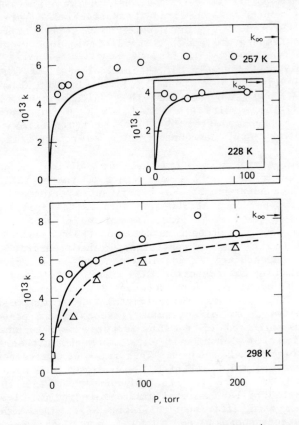

Figure 6. Rate constant data of Ref. 7 (circles) and Ref. 6 (tri-angles) for the OH + C₂H₂ reaction at 228K, 257K, and 298K. The solid line is a fit to the data of Ref. 7 and the dashed line is a fit to the data of Ref. 6. See text for discussion of the fitting procedure. Marked on each plot is the calculated value of k_∞.

by Michael et al. (7). We consider the computed curves to be an excellent representation of the data over this broad range. Also included on Figs. 6 and 7 are the results of Perry and Williamson (8) for two temperatures. Their data are not fit as well, due largely to the points in each case at the two highest pressures which do not conform to the gradual approach to k_∞ predicted by expressions (5) and (7). The single point of Pastrana and Carr (11) at room temperature and 1 torr is within experimental error of the value predicted by the fit.

With this confidence in our theoretical model of the pressure dependence of the addition channel (1) for the $OH + C_2H_2$ reaction up to 420K, we may extend the calculations to higher temperature. The results are shown in Fig. 8 for a pressure of 100 torr, the highest attained in the LP/LF experiments, and a pressure of 760 torr, corresponding to an atmospheric pressure flame.

From Fig. 8 it can be seen that the addition channel may be expected to contribute only a minor part of the overall reaction rate for the conditions of our experiments performed at 1100K. Some pressure dependence at 900K, as seen in the preliminary runs at this lower temperature (5), is also in accord with the calculation. Thus the present results and those measured at lower temperatures (6, 7, 8) are consistent. The predictions for the intermediate temperature region may be used as a guide to the modeller including this reaction in a detailed treatment of flame chemistry. We emphasize, however, that the calculation is not a substitute for accurate experimental determination. The parameters on which it relies, while reasonable, are based more on the high pressure end of the low-temperature data whereas the crucial region for 1 atm and below is much further into the fall-off at higher temperatures. That is, k (1 atm) depends on k_∞ at low temperature and k_0 at high temperature. (We also note the the calculations shown in Fig. 8 correspond to 100 torr N_2 pressure, while the experimental gas mixture includes significant amounts of a more efficient collider, SF_6. Thus this rough prediction corresponds to lower experimental pressure, approximately 50 torr).

Also included on Fig. 8 is the rate constant determination for reaction (2) in a flame by Fenimore and Jones (9). The line drawn in this region for the temperature dependence of the direct reaction (2) corresponds to an Arrhenius form. It has a frequency factor 2×10^{-11} cm^3 sec^{-1} and activation energy of 9 kcal/mole, estimated respectively by the $OH + CH_4$ A-factor and an estimate of ΔH for reaction (2) coupled with a 2 kcal/mole barrier. While this forms a reasonable description of the two experimental results, it would be desirable to measure points intermediate in temperature.

Finally, there are several questions not addressed in the present discussion. One is the zero-pressure intercept at low temperature inferred by Michael et al. (7) and attributed to a

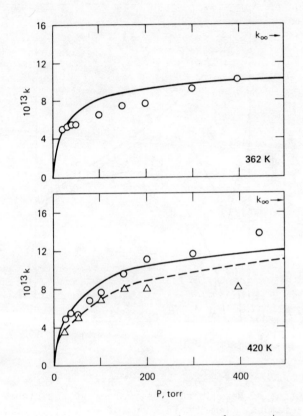

Figure 7. Same as Figure 6 for 362K and 420K.

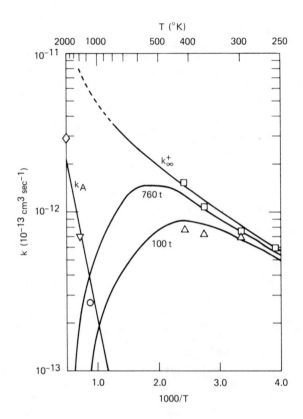

Figure 8. Log k vs. 1/T plot for the OH + C_2H_2 reaction. Triangles, 100 torr values of Ref. 7; squares, values at the highest pressures of Ref. 7; diamond, flame measurement of Ref. 9; inverted triangle, flame measurement of Ref. 10; circle, the average of our twelve experimental points. k_∞^+ line is a transition state theory high-pressure addition rate constant that fits the data of Ref. 7. The 100 torr and 760 torr lines are computational fits to the data of Ref. 7 at these total pressures. The abstraction k_A line is drawn with an A factor of 2 x 10^{-11} cm^3 sec^{-1} and E_a of 9 kcal/mol.

rapid direct rearrangement of the adduct to form ketene and H-atom. Our fit indicates the data of Ref. (7) is consistent with a standard addition process, Reactions (1,3 and 4), but do not demand the total absence of ketene formation. The rate constant at 1 torr (11) suggests this does not occur to an appreciable degree, while molecular beam measurements (12) show ketene to be a direct product at low pressure although in undetermined quantity. Second, questions of thermal decomposition of the stabilized adduct at high temperatures reduce further the net contribution of the addition channel. Lifetimes under 30 μsec are predicted by the model for temperatures above 1000K at 100 torr. Finally the reverse of the direct reaction (2) could affect measurements in the presence of water vapor. These issues are discussed in Ref. (5).

Conclusions

The LP/LF apparatus, developed for the study of bimolecular radical reactions at elevated temperature, has been used for measurements of the OH + C_2H_2 reaction. At 1100K, this reaction is independent of pressure, indicating the primary channel is a direct route rather than addition to form an adduct. A Troe-based calculation has been performed for the addition mechanism, known to dominate at low temperature from the pressure dependence of previously measured rate constants between 230 and 430K.

The results of the calculation show a dramatic change in the fall-off as temperature increases, and that the addition channel is expected to be very slow at 1100K and 100 torr. For use in computer models of combustion systems incorporating detailed chemical kinetics, it is important that this combined pressure-temperature dependence be taken into account. This is all the more essential when a range of T and P may be encountered during a given combustion process.

These results illustrate the desirability of combining experimental results with a theoretical overview in chemical kinetics investigations. While the computations for the intermediate T, P region are not a quantitative substitute for further experiments, they can provide a useful guide for the inclusion of this reaction in combustion chemical networks.

Acknowledgments

We thank David M. Golden for useful discussions and the suggestion that we undertake the calculation of the addition channel. This research was supported by the Department of Energy, Division of Basic Energy Sciences, under Contract No. DE-AC03-81ER10906.

Literature Cited

1. P. W. Fairchild, G. P. Smith, and D. R. Crosley, Nineteenth
    Symposium (International) on Combustion, The Combustion
    Institute, Pittsburgh, 1982, p. 107.
2. J. Troe, J. Phys. Chem. 83, 114 (1979); J. Chem. Phys. 66,
    4758 (1977).
3. L. Seaman, "SRI PUFF 8 Computer Program for One-Dimensional
    Stress Wave Propagation," Final Report, SRI Project 6802,
    Menlo Park, 1978.
4. P. W. Fairchild, G. P. Smith and D. R. Crosley, J. Chem.
    Phys., 79, 1795 (1983).
5. G. P. Smith, P. W. Fairchild and D. R. Crosley, J. Chem.
    Phys., to be published.
6. R. A. Perry, R. Atkinson and J. N. Pitts, Jr., J. Chem.
    Phys., 67, 5577 (1977).
7. J. V. Michael, D. F. Nava, R. P. Borkowski, W. A. Payne and
    L. J. Stief, J. Chem. Phys. 73, 6108 (1980).
8. R. A. Perry and D. Williamson, Chem. Phys. Lett. 93, 331
    (1982).
9. C. P. Fenimore and G. W. Jones, J. Chem. Phys. 41, 1887
    (1964).
10. W. G. Browne, R. G. Porter, J. D. Verlin, and A. H. Clark,
    Twelfth Symposium (International) on Combustion, The
    Combustion Institute, Pittsburgh, 1969, p. 1035.
11. A. Pastrana and R. W. Carr, Jr., Int. J. Chem. Kin. 6, 587
    (1974).
12. J. R. Kanofsky, D. Lucas, F. Pruss and D. Gutman, J. Phys.
    Chem. 78, 311 (1974).

RECEIVED November 30, 1983

Measurement of the $C_2(a^3\Pi_u)$ and $C_2(X^1\Sigma_g^+)$ Disappearance Rates with O_2 from 298 to 1300 Kelvin

STEVEN L. BAUGHCUM and RICHARD C. OLDENBORG

Chemistry Division, Los Alamos National Laboratory, Los Alamos, NM 87545

The disappearance rates of $C_2(a^3\Pi_u$, v=0, 1, and 2) and C_2 ($X^1\Sigma_g^+$, v=0) in the presence of O_2 have been measured over the 298-1300 K temperature range. The C_2 is produced by multiple-photon dissociation of CF_3CCCF_3 at 193 nm and probed by laser-induced fluorescence. The disappearance rate of $C_2(a^3\Pi_u$, v=0) as a function of temperature can be extremely well represented by the Arrhenius expression k(T) = A exp (-E/RT), with A = 1.49 ± 0.03 × 10^{-11} cm^3 molecule^{-1}s^{-1} and E = 0.98 ± 0.02 kcal/mole. The quality of the fit over such a large temperature range provides a test of previously proposed models of the C_2 + O_2 system, which do not predict simple Arrhenius behavior. Our results are consistent with a model in which $^1C_2$ and $^3C_2$ are interconverted by O_2 via a long range interaction and reaction occurs upon still closer approach, so that the identity of the initial state is lost before reaction occurs. Experiments with $C_2(X^1\Sigma_g^+)$ + O_2 are consistent with this model.

Although radical-molecule reactions play an important role in combustion, state-selected measurements of the reaction rate constants at high temperature have been made for only a few reactions. Data on the temperature dependence of radical-molecule reaction rate constants are vital for assessing the importance of various reactions in combustion and similar high temperature processes and provide important insight into the details of the potential energy surfaces involved. Radicals which are electronically or vibrationally excited may react with different rates and produce different products compared to the

0097–6156/84/0249–0257$06.00/0

ground state radical. Thus, it is important to do state selective measurements, where possible.

Detailed reaction mechanisms and kinetics are best studied
in an environment in which specific reactions or sequences of
reactions can be isolated. In order to conduct such studies, we
have constructed a cell in which the temperature can be continuously varied over the 298-1300 K temperature range. Radicals are
produced by photolysis of suitable precursors with a pulsed rare
gas-halide excimer laser and the radical populations probed by
laser-induced fluorescence (LIF). LIF is a powerful tool for
studying kinetic processes, since it is both very sensitive and
state selective, allowing the study of radical-molecule reactions
under pseudo first order conditions. Furthermore, since spatially it is a point diagnostic, high temperature studies can be done
in a relatively small volume, which minimizes some of the experimental difficulties associated with the high temperatures, particularly thermal gradients. A variety of important radicals can
be probed by LIF, including OH, HS, CH, C_2, C_3 and CH_3O.

Our initial experiments have centered on the C_2 radical,
which is known to exist in high concentrations in flame and other
combustion environments. The presence of a low-lying excited
electronic state ($a^3\Pi_u$) within 610 cm^{-1} of the ground state
($X^1\Sigma_g^+$) (1) requires that kinetic studies be done on both states
since both will be significantly populated at the temperatures of
interest. The kinetics of C_2 with a variety of reactants have
been investigated at room temperature (2-7) and over the 300-
600 K temperature range (8-10). We chose to study the C_2 + O_2
reaction as a test of our apparatus and to further evaluate the
model proposed by Mangir and coworkers (6) which predicted non-
Arrhenius behavior of the disappearance rate constants.

Experimental

The high temperature cell (Figure 1) is based on a design by
Felder and coworkers (11). A central high-purity alumina tube is
heated resistively in two zones by Pt/40% Rh resistance wire.
Thermal insulation is provided by an alumina heat shield sur-
rounded by zirconia fiber insulation, with the whole assembly
enclosed in a water-cooled brass vacuum chamber. The temperature
is measured by thermocouples inserted through O-ring seals to
probe various regions of the oven. The thermocouple outputs are
sent to a Micricon microprocessor which automatically regulates
the heater current. The reactant and buffer gases are introduced
at the bottom of the cell and are heated as they flow slowly
(~0.15 sℓm) through the center tube. A small amount of the radical precursor in a helium mixture is introduced via a water-
cooled variable length inlet system within a few cm of the opti-
cal ports to minimize pyrolysis and pre-reaction problems. The
pressure is measured with a capacitance manometer and gas

flows with calibrated Tylan mass flow meters. The gas flow is sufficiently fast to assure a fresh gas mix for each laser shot. At the pressures (20 torr) and flow conditions used, the thermal gradients in the region probed were less than 1%.

The photolysis laser is a Lambda Physik EMG-102 rare gas-halide excimer laser operating at 193 nm. The radical populations are probed using either Quanta-Ray Nd:YAG laser-pumped dye lasers with non-linear mixing crystals where appropriate or a Molectron nitrogen-laser pumped dye laser. The excimer laser beam is focused to a 3 × 3 mm beam and the probe lasers are combined using suitable dichroic mirrors to probe the center of this spot. The fluorescence is imaged with a lens through a suitable dielectric filter onto the element of an RCA 31034A photomultiplier tube. The signal is amplified and processed with a PAR Model 162 boxcar averager. The output of the boxcar is then sent to a computer for sophisticated data analysis. To measure the chemical lifetime of the radical, the time delay between the excimer laser and the probe laser is scanned while monitoring the total fluorescence in the wavelength region of the band of interest. For spectral scans, the time delay is fixed and the dye laser is scanned over the region of interest.

The C_2 radical is produced by multiple-photon dissociation of CF_3CCCF_3 at 193 nm. Approximately 20 torr of helium is used to translationally and rotationally equilibrate the C_2 with the bath gas. The dissociation produces both $^1C_2$ and $^3C_2$ and a significant amount of the $^3C_2$ is vibrationally excited. Our experiments indicate that at these pressures helium is not effective at vibrationally quenching the $^3C_2$, although the rapid rotational thermalization is clearly evident. For measurements of the reaction rates of C_2, 1-2 mtorr of CF_3CCCF_3 is used with 20 torr of 99.99% helium and 0-0.8 torr of 99.99% O_2.

The $C_2(a^3\Pi_u, v=0)$ is probed by excitation of the Swan bands $(d^3\Pi_g \leftarrow a^3\Pi_u)$ on the (0,0) band at 516 nm and the fluorescence from the (0,1) band monitored at 564 nm (1). Excited vibrations are probed either by excitation of the (v,v) band while monitoring the (v,v+1) band or by excitation of the (v-1,v) band with observation of the (v-1,v-1) band fluorescence. $C_2(X^1\Sigma_g^+)$ was initially monitored by excitation of the Phillips band system $(A^1\Pi_u \leftarrow X^1\Sigma_g^+)$ on either the (4,0) band at 691 nm or the (3,0) band at 771 nm while monitoring either the (4,1) fluorescence at 791 nm or the (3,1) fluorescence at 899 nm. Unfortunately, since this transition is relatively weak with a radiative lifetime of 11 μs (12) and in the red, monitoring the $^1C_2$ is very difficult particularly at high temperatures where blackbody radiation provides a significant source of noise in the experiments when monitoring red fluorescence. LIF probes in which the fluorescence in the blue or ultraviolet is monitored are much more useful for high temperature experiments. For $^1C_2$, the Mulliken bands $(D^1\Sigma_u^+ \leftarrow X^1\Sigma_g^+)$ (1) would be ideal but the spectra arising from different vibrational states overlap badly. To surmount these difficulties, we have used double resonance LIF as the probe for $C_2(X^1\Sigma_g^+)$. The (3,0) band of the $A^1\Pi_u \leftarrow X^1\Sigma_g^+$ transition is excited at 771 nm. Fifty nanoseconds later, a second laser excites the (2,3) band of the $C^1\Pi_g \leftarrow A^1\Pi_u$ transition at 404 nm and fluorescence is monitored from the (2,1) band at 359 nm. Since the radiative lifetime of the C state is 32 ns (13) the process is very efficient and a much higher signal-to-noise ratio is obtained than with simple LIF on the Phillips band. Further details of this diagnostic will be presented elsewhere (14).

Results

The time behavior of the $C_2(a^3\Pi_u)$ population under typical conditions is illustrated in Figure 2. The very fast rise occurs when the excimer laser fires and the C_2 is produced. There is some early complicated time behavior followed by a much longer single exponential decay. The analysis of the long time behavior is the goal of this paper. The more complicated analysis of the early time behavior, which depends on the reaction rate, collisional intersystem crossing rate, vibrational energy transfer rates, and the relative $^1C_2/^3C_2$ quantum yields from the dissociation, will be presented in a later paper (14).

A Stern-Volmer plot of the disappearance rate of $C_2(a^3\Pi_u,$ v=0) as a function of oxygen pressure at 886 K is shown in Figure 3. The slope of this plot is the reaction rate constant of $C_2(a^3\Pi_u,$ v=0) with O_2 at 886 K. The quality of the fit gives a good indication of the precision of the measurements. Plots of similar quality were obtained for the other temperatures studied.

A logarithmic plot of the measured rate constants vs $\frac{1}{T}$ for the first three vibrational levels of the $a^3\Pi_u$ state is shown in Figure 4. The straight line through the v=0 points is a fit to the Arrhenius equation, $k(T) = A \exp(-E/RT)$ with A = 1.49 ± 0.03 × 10^{-11} cm^3 molecule^{-1} s^{-1} and E = 0.98 ± 0.02 kcal/mole. As can be seen, the v=0 state is very well represented by the Arrhenius expression. The excited vibrations do not exhibit Arrhenius behavior and the curves drawn through those points are simply for visualization of the data. At room temperature, the excited vibrational levels disappear rapidly upon collisions with oxygen. At higher temperatures, the rates for the v=0, v=1, and v=2 states seem to converge. The detailed analysis of the temporal

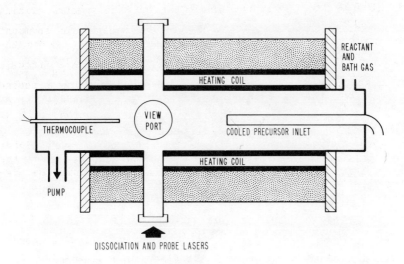

Figure 1.   Diagram of the high-temperature cell.

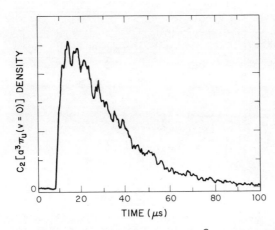

Figure 2.   Typical time behavior of $C_2(a^3\pi_u)$ with O_2 pressure = 0.568 torr, helium pressure = 20.081 torr, temperature = 886 K.

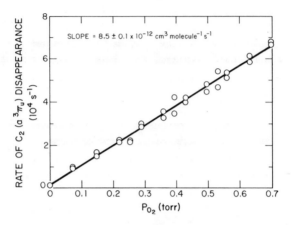

Figure 3.   Stern-Volmer plot of the disappearance rate of
$C_2(a^3\pi_u$, v = 0) as a function of oxygen pressure at a temperature
of 886 K.

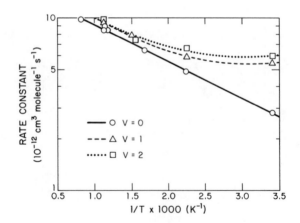

Figure 4.   Log plot of the measured disappearance rate constants
for $C_2(a^3\pi_u$, v = 0, 1, and 2) with O_2 vs. 1/T.

behavior of the v=0 state supports the idea that at room temperature vibrational relaxation of the excited states dominates the kinetics while at higher temperatures the excited vibrational states react with vibrational relaxation playing a more minor role.

The data for the $C_2(X^1\Sigma_g^+, v=0)$ state are presented in Figure 5. The straight line superimposed on the plot corresponds to the results of the Arrhenius fit of the $a^3\Pi_u$ (v=0) data. The measured points fall slightly below the straight line representing the $\tilde{a}$ state data. The reason for this will be discussed later.

Discussion

In the $C_2 + O_2$ system, three basic processes can occur:

$$^1C_2 + O_2 \xrightarrow{k_1} \text{Products} \tag{1}$$

$$^3C_2 + O_2 \xrightarrow{k_3} \text{Products} \tag{2}$$

$$^3C_2 + O \underset{k_e'}{\overset{k_e}{\rightleftharpoons}} {}^1C_2 + O_2 \tag{3}$$

The differential equations for these processes can be solved analytically and the results have been presented by Mangir and co-workers (6). They studied this system at room temperature and showed that the intersystem crossing rate constant, $k_e = 2.7 \times 10^{-11}$ cm^3 molecule^{-1}s^{-1}, was much faster than the observed reaction rate constant. In that case, the equations can be simplified to give

$$k_{obs} = \frac{k_1 + Kk_3}{1 + K} \tag{4}$$

where k_{obs} is the experimentally observed disappearance rate for both $^1C_2$ and $^3C_2$ and K is the equilibrium constant,

$$K = \frac{k_e'}{k_e} = \frac{g_3}{g_1} \exp\left(-\Delta E_{13}/RT\right)$$

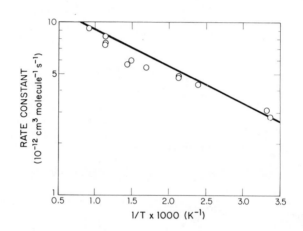

Figure 5. Log plot of the disappearance rate constant for
$C_2(X^1\Sigma_g^+,\ v = 0)$ with O_2 vs. 1/T. The straight line is the result
of the analysis of the Arrhenius fit of the $C_2(a^3\pi_u,\ v = 0)$ data.

ΔE_{13} is the energy separation between the singlet and triplet states, 610 cm^{-1}, and g_3 and g_1 are the degeneracies of the triplet and singlet states, respectively. Over the temperature range of our study, 298-1300 K, the equilibrium constant K varies from 0.32 to 3.06. Thus, the temperature dependence of the observed disappearance rate constant is expected to be non-Arrhenius unless $k_1 = k_3$, which as a first approximation seems unlikely. Previous experiments (10) in the 300-600 K range could not determine if the data were Arrhenius in behavior or not. Our data clearly shows that the observed disappearance rate constant can be fit extremely well to the form

$$k(T) = A \exp (-E/RT) \qquad (5)$$

This result implies that $k_1 = k_3$.

Intersystem crossing between $^1C_2$ and $^3C_2$ by collisions with oxygen is spin allowed and apparently efficient since the singlet-triplet separation is only 610 cm^{-1}. Normally, the chemistry of the singlet and triplet states would be expected to be very different and the apparent result that $k_1 = k_3$ is therefore somewhat surprising. However, if the singlet-triplet crossing occurs at long interaction distances between the C_2 and O_2 and if reaction then occurs upon closer approach, the identity of the initial C_2 state would be lost. Under these circumstances, the experiment would yield $k_1 = k_3$, since the singlet and triplet interactions would be indistinguishable from the kinetic analysis. An analysis of the initial products of the reaction would be useful, but that is beyond the scope of our experiments. A number of very exothermic reaction channels exist (4), but identification of the initial products of the reaction is very difficult. The observed Arrhenius behavior may indicate that the major product channels do not change as the temperature is raised. The observation that vibrational excitation of the $^3C_2$ does not strongly affect the reaction rate may indicate that the C_2 bond is not broken during the reaction, implying C_2O as a product, but with such a low barrier the exit channel of the potential surface may be more important in determining the reaction product.

Our analysis (14) of the temporal behavior of the $^1C_2$ and $^3C_2$ signals at room temperature and at 900 K indicates that the intersystem crossing rate slows down at higher temperature. This detailed modeling indicates that at higher temperatures the $^1C_2$ disappearance rate is affected by the intersystem rate but the $^3C_2$ decay is a single exponential corresponding to the reaction rate. This explains why the $^1C_2$ rate constants show more scatter and lie below the line predicted by the $^3C_2$ result. The details of this are complicated and will be presented later (14).

Conclusions

We have demonstrated that our apparatus can be used to obtain precise kinetic data over the 300-1300 K temperature range. These results with $C_2 + O_2$ suggest that the singlet-triplet intersystem crossing occurs at long range interactions with oxygen and that reaction must occur subsequently upon closer approach. These results clearly demonstrate that kinetic data over a wide temperature range can sometimes provide additional insight into the potential surfaces of the reaction. Future experiments are planned to investigate the reaction kinetics of other radicals.

Acknowledgments

This work was done under the auspices of the U. S. Department of Energy.

Literature Cited

1. K. P. Huber and G. Herzberg, Molecular Spectra and Molecular Structure IV. Constants of Diatomic Molecules; Van Nostrand Reinhold: New York, 1979.
2. H. Reisler, M. Mangir, and C. Wittig, J. Chem. Phys. 1979, 71, 2109.
3. V. M. Donnelly and L. Pasternack, Chem. Phys. 1979, 39, 427.
4. L. Pasternack and J. R. McDonald, Chem. Phys. 1979, 43, 173.
5. H. Reisler, M. Mangir, and C. Wittig, Chem. Phys. 1980, 47, 49.
6. M. S. Mangir, H. Reisler, and C. Wittig, J. Chem. Phys. 1980, 73, 829.
7. H. Reisler, M. S. Mangir, and C. Wittig, J. Chem. Phys. 1980, 73, 2280.
8. L. Pasternack, A. P. Baronavski, and J. R. McDonald, J. Chem. Phys. 1980, 73, 3508.
9. L. Pasternack, W. M. Pitts, and J. R. McDonald, Chem. Phys. 1981, 57, 19.
10. W. M. Pitts, L. Pasternack, and J. R. McDonald, Chem. Phys. 1982, 68, 417.
11. W. Felder, A. Fontijn, H. N. Volltrauer, and D. R. Voorhees, Rev. Sci. Instrum. 1980, 51, 195.
12. P. Erman, D. L. Lambert, M. Larsson, and B. Mannfors, Astrophys. J. 1982, 253, 983.
13. L. Curtis, B. Engman and P. Erman, Physica Scripta 1976, 13, 270.
14. S. L. Baughcum and R. C. Oldenborg, to be published.

RECEIVED October 28, 1983

Reaction of Carbon Monoxide with Oxygen Atoms from the Thermal Decomposition of Ozone
Effect of Added Gases

SIDNEY TOBY, SHAILESH SHETH, and FRINA S. TOBY

Rutgers, The State University of New Jersey, Department of Chemistry, New Brunswick, NJ 08903

The reaction between carbon monoxide and oxygen atoms produced by the thermal decomposition of ozone was studied in the range 80–160°C. The chemiluminescence from $CO_2(^1B_2)$ was used to follow the course of the reaction. The effect of added carbon dioxide, tetrafluoromethane and oxygen on the kinetics and chemiluminescence was investigated. It is concluded that there are simultaneous bimolecular and third body channels for the reaction of CO with O-atoms to produce electronically excited CO_2.

The reaction of oxygen atoms with carbon monoxide is an important reaction in many combustion systems. Although there is an extensive literature on this reaction (1) there is disagreement and uncertainty on the molecularity of the reaction, on the kinetic parameters and on the mechanism of the chemiluminescence. We have investigated this reaction using O-atoms from the thermal decomposition of ozone. This has advantages compared to systems where O-atoms are produced by a discharge through molecular oxygen. We have shown that this is a feasible system to study the reaction $O + CO \rightarrow$ providing that trace impurities are carefully removed from the CO (2). We also described how kinetic data could be obtained from the chemiluminescence. The present work extends this approach and investigates the effect of added gases on the emitted intensity so as to provide more information on the molecularity of the reaction.

Experimental Section

The experimental system has been described previously (2, 3). A cylindrical quartz reaction vessel of volume 0.525L was situated in a thermostatted oven and the ozone concentration was measured by absorption at 254nm. Other gas concentrations were measured by

0097–6156/84/0249–0267$06.00/0

diaphragm gauges and a MKS Baratron capacitance manometer. Chemiluminescence was measured by shuttering the monitoring beam and
measuring the unfiltered emission with a 1P28 photomultiplier.
Ozone was made by passing oxygen (Matheson Ultrapure grade)
through an ozonizer and performing several freeze-thaw cycles so
that the O_2 could be pumped away. Carbon monoxide was used both
from Matheson Research Grade Pyrex bulbs and from commercial cy-
linders after being distilled at -196°. In both cases metal carbonyls are present as impurities and were removed by the method
descrobed by Stedman et al (4) using an iodine/charcoal spiral.
Carbon dioxide and tetrafluoromethane were distilled from the
commercial materials.

Computer simulation of the kinetic mechanism was carried out
using the program of Brown (5) on a Digital Equipment Corp. VAX
780 computer.

Results

As noted previously (2) runs carried out in the absence of a mod-
erator gas tended to be erratic, probably because of a vibrationally excited chain propagator. Experiments were therefore
carried out in the presence of added CO_2, O_2or CF_4.

Ozone decay plots were pseudo first order with correlation
coefficients >0.99 and typical examples are shown in Figure 1.
The pseudo first order constants were divided by the total concentration using efficiencies given by Hampson (6) and the resulting
second order constants are given as k_α in Table I with the experimental conditions. Also given in Table I are the relative emitted
intensities measured by the unfiltered photomultiplier. It was
difficult to measure intensity at the beginning of the experiments
and the parameter $I_{1/2}$, the intensity when $[O_3] = [O_3]_0/2$, is used.

Discussion

When sufficiently purified CO is reacted with O_3 the rate law
reduces to the simple first order dependence expected from the
sequence.

$$O_3 + M \underset{-1}{\overset{1}{\rightleftharpoons}} O_2 + O + M$$

$$O + O_3 \overset{2}{\rightarrow} 2O_2$$

Under the conditions of our experiments $k_2[O_3] \gg k_{-1}[O_2][M]$ for
virtually the entire run and assuming a steady state in [O] yields
$-d[O_3]/dt = 2k_1[O_3][M]$ and this pseudo first-order dependence is
seen in Figure 1. The literature values (6) for $2k_1$ (M=CO_2 or

Table I. Chemiluminescent Intensities and Rate Constants from the Reaction of 0-atoms with CO in the Presence of Added Gases

$P(O_3)$	$P(CO)$	$P(CO_2)$	$P(X)$	$I_{1/2}$,nA		k_a,$\underline{M}^{-1}s^{-1}$
T	o	r	r			
			80°C			
0.25	8.08	0.8		0.75		1.43
0.25	7.35	5.0		0.5		0.73
0.25	7.40	23.0		0.7		0.29
			115°C			
0.25	5.76	1.0		2.2		3.05
0.25	5.26	1.0		1.9		3.24
0.25	5.74	1.0		5.1		6.34
0.25	5.68	1.0		3.5		4.85
			150°C			
0.25	1.2		8.50[a]	6.1		15.0
0.25	1.2		8.48	7.4		15.9
0.25	1.2		4.00	6.9		29.2
0.25	1.2		4.0	6.2		25.5
0.25	1.2		21.7	7.3		6.29
0.25	1.2		1.51	5.25		30.2
0.25	1.2		0.76	5.65		52.2
0.25	1.2		0.76	6.3		54.3
0.25	1.2		0.19	6.0		46.9
0.25	1.2		1.56	6.05		39.7
0.25	1.2		1.38	5.6		36.6
0.25	1.2		4.24	4.8		17.0
0.25	1.2		21.3	5.9		4.10
			160°C			
0.20	0.64	2.5		1.8		14.7
0.20	0.64	2.5		1.65		10.4
0.25	1.06	23.0		9.6		7.47
0.40	1.06	5.0		4.1		8.51
0.20	1.06	16.0		7.65		8.07
0.25	1.00	5.0		4.0		10.2
0.25	3.90	10.0		23.5		22.6
0.25	2.77	10.0		14.2		13.7
0.41	0.72	3.65	5.34[b]	3.42		8.68
0.32	0.87	4.41	2.51	3.70		11.5
0.46	0.59	3.00	8.20	2.88		5.29
0.42	0.73	3.72	5.34	3.44		7.34
0.22	0.65	2.96	8.50	2.60	1.29	4.83
0.24	0.86	4.40	2.60	3.22	1.58	6.70
0.24	0.54	2.75	9.50	2.44	1.30	4.25

(a) At 150° $X \equiv CF_4$ (b) At 160° $X \equiv O_2$

equivalent) at 80, 115, 150 and 160° are 0.017, 0.32, 3.7 and 6.8 $\underline{M}^{-1}s^{-1}$ respectively. These values are approached at the higher values of added gas and this is well illustrated in Figure 2 where the observed second order constants are plotted against the pressure of added CF_4 from the data in Table I. Clearly the thermal decomposition of ozone is approached at sufficiently high pressures of deactivating gas.

Previous simulation work (2) showed that the kinetics of the reaction of O_3 with impure CO could be described using an initiation step O_3 + HX → where HX is a hydrogen-containing impurity. An equally good and more realistic simulation was obtained by postulating a nickel carbonyl impurity as the initiator and using the mechanism given by Stedman and Branch (4) for the reaction of O_3 with nickel carbonyl. Of particular concern to the present work is the fact that the reaction of O_3 with carbonyls is chemiluminescent from excited metal oxides and this emission could interfere from that arising from O + CO → hν. We sometimes found an erratic emission at the beginning of the experiments which then became reproducible as the run proceeded. Any emission from O_3 + Ni(CO)$_4$ → hν would follow the rate law $\underline{I}$ = k [O_3] [Ni(CO)$_4$] and would be expected to decay rapidly because of the small initial concentration of carbonyl (<1ppb (4)). On the other hand, emission from O + CO → hν shows a relatively small time dependence when the 0-atoms are formed from O_3 (2) and thus use of the $I_{1/2}$ parameter should eliminate any interference from carbonyl emission.

The Mechanism

We postulate steps 1, -1, 2 and the following sequence which differs from that previously proposed (2) by the addition of step 8:

$$O + CO \quad \xrightarrow{\ 3\ } \quad CO_2^*(^3B_2)$$

$$CO_2^*(^3B_2) \quad \xrightarrow{\ 4\ } \quad CO_2(^1B_2)$$

$$CO_2^*(^3B_2) + M \quad \xrightarrow{\ 5\ } \quad CO_2 + M$$

$$CO_2(^1B_2) \quad \xrightarrow{\ 6\ } \quad CO_2 + h\nu$$

$$CO_2(^1B_2) + M \quad \xrightarrow{\ 7\ } \quad CO_2 + M$$

$$O + CO + M \quad \xrightarrow{\ 8\ } \quad CO_2(^1B_2) + M$$

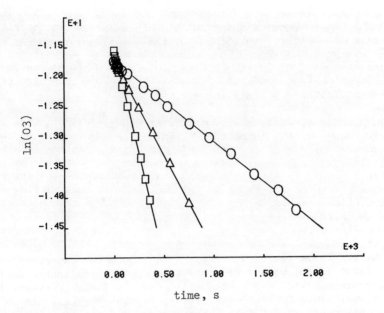

Figure 1. Typical ozone decay plots at 160°. $P(CO) = 1.06$ torr, $P(O_3)_0 = 0.25$ torr, $P(CO_2) = 2.5$ (circles), 100 (triangles), and 23 (squares).

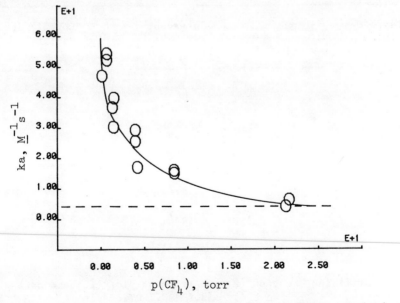

Figure 2. Effect of added CF_4 on measured ozone decay constant at 150°. The dashed line is the literature value of $2k_1$.

Taking steady states in intermediate concentrations (which was
verified by computer simulation) and since nearly all the 0-atoms
disappear via step 2 we obtain for the emission intensity:

$$\frac{I}{} = \frac{k_1 k_6 [CO] [M] (k_3 k_4 + k_8 [M] (k_4 + k_5 [M]))}{k_2 (k_4 + k_5 [M]) (k_6 + k_7 [M])} \quad (1)$$

Equation (1) predicts that $\underline{I} \propto [CO]$ at constant [M] as previously
found (2). We now assume that $k_7[M] \gg k_6$ and that $k_5[M] \ll k_4$
to obtain:

$$\frac{I}{[CO]} = \frac{k_1 k_6 (k_3 + k_8 [M])}{k_2 k_7} \quad (2)$$

Equation (2) is tested in Figure 3 and shows the expected linearity
of $\underline{I}/[CO]$ in [M] with slopes and intercepts which increase with
temperature. Figure 3 contains the data in Table I (except for the
runs with added CF_4) and data from previous work (2). The slopes
and intercepts from Figure 3 are listed in Table II and Figure 4.

Table II. Slopes and Intercepts from Figure 3.

Temp °C	Slope nA $\underline{M}^{-2} \times 10^{-8}$	Intercept nA $\underline{M}^{-1} \times 10^{-4}$	Correlation coefft.
80	–	0.20	–
115	–	1.6	–
132	0.46	1.5	0.943
150	1.2	4.2	0.952
160	2.4	4.7	0.988

Utilizing Equation (2) an Arrhenius plot of these data gives
$\ln(k_1 k_6 k_8 k_2^{-1} k_7^{-1}/nA \underline{M}^{-2}) = 42.7 - 20.2/RT$ and ln

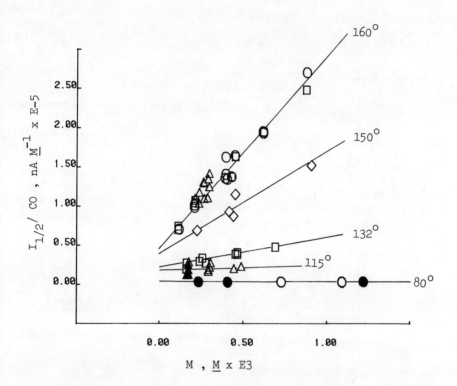

Figure 3.   Plot of $I_{1/2}/(CO)$ vs. (M.) $160°$, triangles, O_2 added.
Data from Ref. 2 are at $160°$ (squares), $150°$ (diamonds), $132°$
(squares), $115°$ (unfilled triangles), and $80°$ (unfilled circles).

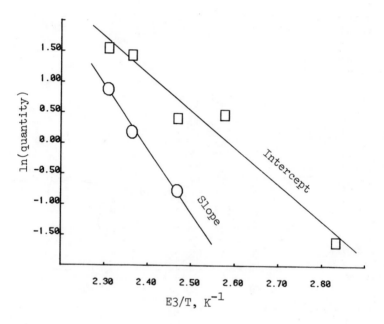

Figure 4. Arrhenius plot of slopes and intercepts from Figure 3.

$(k_1 k_3 k_6 k_2^{-1} k_7^{-1}/nA \underline{M}^{-1}) = 24.6-11.8/RT$ where R is in kcal mole^{-1} K^{-1}. We cannot obtain absolute values of $\underline{I}$ which is dependent on the geometry of our system, but combining the Arrhenius lines gives with estimated errors $\ln (k_8 k_3^{-1}/\underline{M}^{-1}) = 18.1 \pm 2.2 -(8.4 \mp 1.1)/RT$. Combining this with the value of $\overline{k}_3$ obtained previously (2) gives $\ln (k_8/\underline{M}^{-2} s^{-1}) = 33.7 \pm 2.2 -(11.6 \mp 1.1)/RT$. Termolecular reactions involving 0-atoms have negative temperature coefficients except for the reactions with N_2, SO_2, and CO which have positive activation energies. We may compare reaction 8 with the isoelectronic reaction $0 + N_2 + M \longrightarrow N_2O + M$ for which the reported activation energy is 10.4 kcal mole^{-1} (6).

A potential energy diagram for the lower excited states of CO_2 has been given by Pravilov and Smirnova (7) who have summarized the recent literature on the reaction of 0-atoms with CO. They have pointed out the complications resulting from impurities and heterogeneous effects on this system.

We conclude that the thermal decomposition of O_3 is a useful source of 0-atoms providing that O_3 does not react appreciably with the substrate. In the case of the reaction with CO traces of carbonyl impurity complicate the kinetics and chemiluminescence, especially at the lower temperatures. These complexities can be overcome and our results indicate that there are simultaneous bimolecular and third body channels for the reaction of 0-atoms with CO. This may account for the discordant literature for this reaction.

Acknowledgment

We thank the Research Council of Rutgers University for support of this work.

Literature Cited

1. Dixon-Lewis, G.; Williams, D.J. in "Comprehensive Chemical Kinetics" Vol. 17. Bamford, C.M.; Tipper, C.F., Eds.; Elsevier: The Netherlands, 1977.
2. Toby, S.; Sheth, S.; Toby, F.S. Int. J. Chem. Kinetics 1983, 15, 0000.
3. Toby, S.; Ullrich, E. Int. J. Chem. Kinetics 1980, 12, 535.
4. Stedman, D.H.; Tammaro, D.A.; Branch, D.K.; Pearson, R. Analyt. Chem. 1979, 51, 2340.
5. Brown, R.L. "A Computer Program for Solving Systems of Chemical Rate Equations", NBSIR 76-1055, Natl. Bur. of Standards, Washington, D.C., 1978.

6. Hampson, R.F. "Chemical Kinetic and Photochemical Data Sheets
    for Atmospheric Reactions", FAS-EE-80-17, Natl. Bur. of
    Standards, Washington, D.C. 1980.
7. Pravilov, A.M.; Smirnova, L.G. Kinetika i Kataliz 1981, 22,
    832.

RECEIVED November 29, 1983

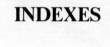

INDEXES

Author Index

Subject Index

A

Production by Paula Bérard
Indexing by Florence Edwards
Jacket design by Anne G. Bigler

Elements typeset by Hot Type Ltd., Washington, D.C.
Printed and bound by Maple Press Co., York, Pa.

RECENT ACS BOOKS

"Geochemical Behavior of Disposed Radioactive Waste"
Edited by G. Scott Barney, James D. Navratil, and W. W. Schulz
ACS SYMPOSIUM SERIES 248; 470 pp.; ISBN 0-8412-0831-X

"NMR and Macromolecules:
Sequence, Dynamic, and Domain Structure"
Edited by James C. Randall
ACS SYMPOSIUM SERIES 247; 282 pp.; ISBN 0-8412-0829-8

"Geochemical Behavior of Disposed Radioactive Waste"
Edited by G. Scott Barney, James D. Navratil, and W. W. Schulz
ACS SYMPOSIUM SERIES 246; 413 pp.; ISBN 0-8412-0827-1

"Size Exclusion Chromatography: Methodology and
Characterization of Polymers and Related Materials"
Edited by Theodore Provder
ACS SYMPOSIUM SERIES 245; 392 pp.; ISBN 0-8412-0826-3

"Industrial-Academic Interfacing"
Edited by Dennis J. Runser
ACS SYMPOSIUM SERIES 244; 176 pp.; ISBN 0-8412-0825-5

"Characterization of Highly Cross-linked Polymers"
Edited by S. S. Labana and Ray A. Dickie
ACS SYMPOSIUM SERIES 243; 324 pp.; ISBN 0-8412-0824-9

"Polymers in Electronics"
Edited by Theodore Davidson
ACS SYMPOSIUM SERIES 242; 584 pp.; ISBN 0-8412-0823-9

"Radionuclide Generators: New Systems
for Nuclear Medicine Applications"
Edited by F. F. Knapp, Jr., and Thomas A. Butler
ACS SYMPOSIUM SERIES 241; 240 pp.; ISBN 0-8412-0822-0

"Polymer Adsorption and Dispersion Stability"
Edited by E. D. Goddard and B. Vincent
ACS SYMPOSIUM SERIES 240; 477 pp.; ISBN 0-8412-0820-4

"Assessment and Management of Chemical Risks"
Edited by Joseph V. Rodricks and Robert C. Tardiff
ACS SYMPOSIUM SERIES 239; 192 pp.; ISBN 0-8412-0821-2

"Archaeological Chemistry--III"
Edited by Joseph B. Lambert
ADVANCES IN CHEMISTRY SERIES 205; 324 pp.; ISBN 0-8412-0767-4

"Molecular-Based Study of Fluids"
Edited by J. M. Haile and G. A. Mansoori
ADVANCES IN CHEMISTRY SERIES 204; 524 pp.; ISBN 0-8412-0720-8